PENGUIN

THE PENGUIN BOOK OF
OUTER SPACE EXPLORATION

JOHN LOGSDON, known as the "dean of space policy," is one of America's foremost experts on space policy and history. In 1987 he founded the Space Policy Institute at George Washington University, which he directed until 2008. He is a former member of the NASA Advisory Council, and he served on the Columbia Accident Investigation Board in 2003. He is the editor of a seven-volume series, *Exploring the Unknown: Selected Documents in the History of the U.S. Civil Space Program*, and his books include *John F. Kennedy and the Race to the Moon, After Apollo?: Richard Nixon and the American Space Program*, and the forthcoming *Ronald Reagan and the Space Frontier.* He serves on the board of directors for The Planetary Society, as well as several other prestigious space-related organizations and journals.

BILL NYE is best known as the host of PBS's *Bill Nye the Science Guy.* He is the bestselling author of several books, including *Everything All at Once*, and the host of Netflix's *Bill Nye Saves the World.* He is also CEO of The Planetary Society, the world's largest nonprofit space organization.

The Penguin Book of Outer Space Exploration

NASA and the Incredible Story of Human Spaceflight

Edited by
JOHN LOGSDON

Foreword by
BILL NYE

PENGUIN BOOKS

PENGUIN BOOKS

An imprint of Penguin Random House LLC
375 Hudson Street
New York, New York 10014
penguinrandomhouse.com

LIBRARY OF CONGRESS CATALOGING-IN-PUBLICATION DATA
Names: Logsdon, John M., 1937– editor. | Nye, Bill writer of foreword.
Title: The Penguin book of outer space exploration : NASA and the incredible
story of human spaceflight / edited by John Logsdon ; foreword by Bill Nye.
Other titles: NASA and the incredible story of human spaceflight
Description: First edition. | New York, New York : Penguin Books, [2018] |
Includes bibliographical references.
Identifiers: LCCN 2018009281 (print) | LCCN 2018012899 (ebook) |
ISBN 9781101993491 (Ebook) | ISBN 9780143129950 (pbk.)
Subjects: LCSH: Manned space flight—United States—History. | United States.
National Aeronautics and Space Administration—History. | Outer
space—Exploration—History.
Classification: LCC TL873 (ebook) | LCC TL873 .P46 2018 (print) |
DDC 629.450973—dc23
LC record available at https://lccn.loc.gov/2018009281

Printed in the United States of America
1   3   5   7   9   10   8   6   4   2

Set in Sabon LT Std

Contents

Foreword

There have been hundreds of thousands of documents written for and about the U.S. space program, literally tons. They tell a story shaped not only by domestic politics and international political and military conflict but also by the strange desire deep within us to do and achieve something greater than ourselves—to glean something of the grand mystery of creation. That story is all here, in more than one hundred documents that shaped space history. With these few small steps, humankind became a spacefaring species. Many more steps are to come.

It might seem obvious looking back, but when spaceflight began, it wasn't clear which team—by that I mean which space program, sponsored by which government in which country—would make it to the Moon first, and it certainly wasn't obvious how profoundly the world would be changed by humanity's ventures into space. In this book, John Logsdon uses the original source materials to take us on a journey from the shores of Florida to the farthest reaches of the cosmos. Did you know that prior to Yuri Gagarin's flight, there were fears that a human simply could not survive being launched into space? Did you know the first guys to walk on the Moon filled out customs forms? After all, they had been out of the country for a few days. Did you know that President Nixon used the shuttle program to ensure votes for his reelection? Did you know that in 1993 it was Russian space leaders who suggested merging the U.S. and Russian space station programs? The evidence is here, in these documents.

The story of space exploration is marked by a series of turning points, a series of policy decisions. These decisions are documented well enough, but the key documents would be *very* hard to find, if you didn't know where to look. Dr. Logsdon does. He's the dean of space history. He is the world's foremost authority on which of the hundreds of thousands of documents hold the keys to knowing what and why significant things happened on Earth that influenced our presence in space. For many of us, the most significant event in the history of space exploration to date remains the first human landing on the Moon, which Neil Armstrong called "one small step for a man, one giant leap for mankind." But in my view, there is little doubt that the next grand achievement in space will render the Moon landings both a giant leap for their time and but a small step toward the future.

The documents included here show that once the Soviet Union had sent its robotic emissary, the first Sputnik satellite, to the ultimate high ground of space, the world changed. Humans vied for access to and control over what would be a limitless frontier. The United States created the National Aeronautics and Space Administration (NASA), and the space race was born. From the beginning, there was conflict and tension between those in presidential administrations who wanted to achieve strategic or geopolitical objectives in space and those who wanted to explore space for the sake of science. As today's commercial space companies grow, this tension will manifest itself in new ways. Here's hoping these documents will help us find the best course forward.

At the height of the competition to get to the Moon, NASA's budget was ten times what it is today. It led to the return of the famed Moon rocks, which solidified our understanding of the age of the Moon, the age of the Earth, and the Moon's origin. The astronauts' gloves-on exploration, along with all sorts of other geologic evidence, led researchers to conclude that the Moon is made of a large ball of earth—from the planet Earth. Our Moon is made of once-molten Earth's crust. It is a profound insight. It gets at an answer to the age-old question: Where did we all come from?

However, for me it frankly pales compared with a single photograph that was taken before anyone set moonboot on the lunar surface. That Apollo 8 image has come to be called the "Earthrise" photo. That image changed the way everyone on Earth viewed our place in the blackness of space. Documents included here show how the decision to send Apollo 8 to the Moon was made. But did we need a human explorer, who wasn't even a professional photographer, to take that picture? Did we really need to send people there to collect rocks? The Soviet program did the same thing(s) with a series of robotic explorers. While the Soviet missions to the Moon are seldom remembered, the twelve Americans who walked on the lunar surface deserve their place in human history. As one 1961 document included here says, "It is man, not merely machines, in space that captures the imagination of the world." On the other hand, our cosmic view has been enhanced immeasurably by the amazing images taken by robotic spacecraft silently hurtling through the cold dark and from above and upon the surfaces of other worlds. But what will our view of the cosmos be like after human explorers have set boots on the soil of Mars? Will we remember the noble robots sent before? Read on, and judge for yourself.

After great achievements on the Moon, the U.S. space program was pulled back. Rather than driving farther and deeper into space, the program and its mission were redirected to more Earthly pursuits, or more accurately, more just-above-the-Earth's-atmosphere pursuits. The shuttle program kept thousands of space professionals engaged and created a fleet of spaceships that proved to be expensive and hard to operate safely. While hundreds of astronauts traveled through the hard vacuum of low Earth orbit, not much new was accomplished scientifically, with the important exception of the Hubble Space Telescope, which is still in service as I write. Hubble has helped us understand the origin of the universe, the Big Bang, the objects beyond Neptune and Pluto. It has given us thousands of images that are easily recognized by humans the world around. Hubble was launched with a flawed mirror, and only because it was designed to be serviced by astronauts was it able to deliver on its promise.

Some question what the United States and other nations would have achieved if the same resources directed toward competition in space had instead been brought to bear on technical problems down here on Earth. I believe, if you study the history presented here, you'll see it's not a fair question. The records show that NASA was created to compete, and the competition of the Cold War led to extraordinary technologies. We have handheld devices that provide five or more independent systems of communication, remarkably accurate navigation, and endlessly varied software applications that can notify us of emergencies, track our health, and entertain us. It's all been brought down to Earth from space.

It may surprise you to read that in addition to science, the shuttle program and the International Space Station have contributed a great deal to diplomacy. In their native countries, there are no bigger heroes than the astronauts—from Canada, Italy, Japan, Sweden, and the United Kingdom, among many other countries— who've flown a few hundred nautical miles straight up from home. The legacy of this part of the U.S. space program is more political than technical, but nonetheless very real and extremely significant.

Not long before NASA was redirected to fly primarily in low Earth orbit, other workers in the agency and the scientific community wanted to see what was out there, farther from home, and far more difficult to explore. There was in the 1960s a concurrent push for exploring Mars. The first Mars probe was a repurposed and modified version of the Ranger spacecraft that had been designed and flown to the Moon. (By flown to the Moon, we mean purposefully smash-landed on the Moon.) As Mariner 4, the same design sent us the first pictures of another world. This led scientists and engineers to propose and succeed in flying two spacecraft that soft-landed on Mars. The first pictures went around the world at the speed of light. For the first time, humankind saw that our planet had a neighbor, a kindred world with a rocky surface that looks not altogether different from areas here on our home world. The astonishing nature of the Martian landscape has motivated space agencies around the world, NASA especially, to explore Mars and other planets in our solar system with rigor—and to set as a "horizon goal" eventual human travel to the surface of the Red Planet, a vision articulated by Wernher von Braun more than six decades ago and echoed today by such visionaries as Elon Musk.

Today in space—and here on Earth—we find ourselves wrestling with the same problems that have faced policy makers since the beginning of the race to space. We have military spacecraft by the hundreds. We have weather satellites by the dozens. We have lawmakers vying for position to get budget dollars directed to their districts to keep space hardware in the works and get it up to orbit. But now, unlike any time in space history, we have commercial interests that are planning to explore by making sales, not to governments that seek capable rocket boosters, capsules, and support equipment, but to citizens. Companies like Blue Origin, Sierra Nevada Corporation, SpaceX, and Virgin Galactic want to send people and supplies to space, not for science, but for exploration and eventual habitation.

When we explore, two things happen. We make discoveries, to be sure. But perhaps far more important, we have adventures. It's the promise of adventure that drives these companies, their leaders, and their personnel. As you read this volume, I hope you'll feel the tension, and I hope you'll think hard about what is the best use of our intellect and treasure. Do we explore for exploration's sake? Do we operate up there for the communication, surveillance, and forecasting capabilities that spacecraft can provide? Is space for the military more than anything? A consensus between government space agencies, companies selling hardware and expertise to the agencies, and the aspirations of the leaders of these extraordinary emerging companies is in the offing. Here's hoping historical insights drawn from these pages will help all the parties involved seek the best way forward.

For me, and I hope for you, the promise of a discovery on another world makes all that we do for the sake of space—the technology, the employment, the pure science—worth it. It would all be worth every penny, peso, euro, yen, yuan, rupee, and ruble, with every sacrifice, and every hour spent in meetings with middle managers, if we were to discover evidence that life once existed on another world. Stranger still, we may discover something still alive on Mars or Europa (the moon of Jupiter with twice as much seawater as Earth). It would change the way each and every one of us views our relationship to the cosmos. As citizens of the Earth, when these discoveries are made, you and I would be part of history. We would have supported some of the intellect and treasure that would make a discovery akin to that of Copernicus, Newton, or Galileo. It would be profound.

To understand how we all got on this path to discovery, open these pages and let Dr. Logsdon show you the future through study and understanding of space history, a history that has to date unfolded in a few small steps.

BILL NYE

Introduction

This volume, which aims to tell the story of America's journey to space through the words of those involved in this historic undertaking, reflects my more than half-century involvement in the history of the U.S. space program. Let me start with the most memorable day of that half century.

Early in the morning of July 16, 1969, I was among a few hundred people at Kennedy Space Center's Operations and Checkout Building waiting for space-suited Neil Armstrong, Buzz Aldrin, and Michael Collins to walk by us—on their way to the Moon! A few hours later, precisely at 9:32 a.m., I stood in the field fronting the iconic KSC countdown clock and watched the three explorers lift off atop the powerful Saturn V booster. I knew then I was witnessing history being made, and I had already begun to play a role in recording it.

The reason I had the unforgettable experience of being present as humans for the first time began the voyage to land on another celestial body was that I had chosen the 1961 decision by President John F. Kennedy to send Americans to the Moon and return them safely to Earth "before this decade is out" as the topic for my doctoral dissertation in political science. By mid-July 1969 the dissertation was essentially complete and was already close to acceptance for publication by the MIT Press as *The Decision to Go to the Moon: Project Apollo and the National Interest* (1970). I had worked closely with the NASA History Office in preparing the study, and was rewarded with an invitation to attend the launch.

The dissertation was the culmination of my graduate studies at New York University. Throughout those studies, I had centered my research papers on space-related topics whenever possible, because the U.S. space effort of the 1960s incorporated every aspect of the international relations and foreign policy focus of my graduate program. I had become fascinated by the space program after watching John Glenn parade through the streets of Manhattan on March 1, 1962, soon after becoming the first American to orbit the Earth. To me, the early voyages into orbit during the Mercury and Gemini programs were the most exciting things happening, and I closely followed the progress—and setbacks—of the Apollo program as it prepared for humanity's first exploratory voyages across what John Kennedy called "this new ocean" of space.

I have chosen to begin this volume on such a personal note because its contents reflect my continuing fascination with space activities, and particularly human spaceflight. I could not have anticipated as Apollo 11 accelerated off Pad 39A that July morning that I would devote a long career to the study of space policy and history, but that is indeed what happened.

In 1987 I founded the Space Policy Institute as a research and graduate education unit of George Washington University's Elliott School of International Affairs. One of the first large research contracts for the Space Policy Institute came from the NASA History Office. NASA chief historian Sylvia Kraemer had the idea of creating a reference work containing documents seminal to the evolution of the U.S. civil space program, and in 1989 the Space Policy Institute won the competition to prepare that documentary history, with me as lead editor. The original notion was that the project would result in two volumes. It took a while to assemble the team to prepare the initial volume, and in fact it was not published until 1995. By that time, Dr. Kraemer had moved out of the chief historian position and was replaced by Roger Launius. It was also clear that two hefty volumes were insufficient to encompass all aspects of space program history; working with Dr. Launius and his successor, Steve Dick, the Space Policy Institute eventually produced the seven-volume series *Exploring the Unknown: Selected Documents in the History of the U.S. Civil Space Program*. The seven volumes represent the work of many contributors at both the Space Policy Institute and NASA; they are acknowledged in the individual volumes and are too numerous to be noted again here. These seven volumes, which are too extensive for the casual reader, form the basis of this one-volume Penguin Classics edition. They can be accessed at https://history.nasa.gov/series95.html.

With the 1958 creation of NASA and the 1960 issuance of a National Space Policy, the Eisenhower administration created the framework for a U.S. civilian space program carrying out both robotic and human missions of discovery and exploration. But, as noted in the 1960 policy, "manned space flight and explora-tion will represent the true conquest of outer space." This collec-tion of documents reflects that observation, focusing on the U.S. program of human spaceflight, from its first tentative steps off the planet, through the twelve U.S. astronauts who walked on the surface of the Moon, and then the half century and still counting period during which human spaceflight has been lim-ited to the near vicinity of Earth, to the plans for once again sending humans into deep space, perhaps first back to the Moon and eventually to the surface of Mars. The goal of this account is to trace the "few small steps" that have propelled Americans off their home planet.

In their May 8, 1961, memorandum recommending that a lunar landing be set as a U.S. national goal, NASA administrator James Webb and Secretary of Defense Robert McNamara ob-served that it is "men, not merely machines, in space that capture the imagination of the world." That observation, in my view, re-mains valid; just over 560 humans have traveled to orbit and be-yond, and their experiences still represent for many the most exciting aspects of space activity. Certainly robotic spacecraft ex-ploring the solar system and probing the mysteries of the cosmos have produced remarkable images as well as path-breaking sci-ence, but identifying with those people who have actually made the journey themselves is a persistent phenomenon.

The majority of documents in this volume are drawn from the *Exploring the Unknown* series, particularly Volume I, "Organiz-ing for Exploration," and Volume VII, "Human Spaceflight: Projects Mercury, Gemini, and Apollo." In many cases, I have deleted portions of the documents for the sake of keeping this volume within a manageable size. These deletions are indicated by a " * * *" mark. In some cases, small deletions are marked with ellipses. The NASA series did not include material easily ac-cessible elsewhere, but I have added to this volume key presiden-tial speeches and statements, available in the *Public Papers of the*

Presidents of the United States. I have also added, particularly in chapter 4 of this volume, documents related to human spaceflight since the end of the Apollo program. Some of this material comes from my 2015 book, *After Apollo?: Richard Nixon and the American Space Program* (Palgrave Macmillan); another source is my ongoing research on space policy decisions during the past thirty-five years.

I have not provided source information for the documents included, on the grounds that this volume is intended not as a reference work but rather as a cumulative story of the U.S. movement into space as captured in primary documents. It is fascinating to me, and I hope it will be to the reader, to address directly the thinking of those who decided what the United States should do in space and of those who carried out those decisions.

There are many rationales for going into space, ranging from scientific discovery, international competition, national security, national power, and national pride, to commercial profit and societal benefits. All of those rationales are reflected in these documents and have influenced the U.S. space program from its inception more than a half century ago. Underpinning them, I believe, has been a vision of the United States taking a leading role in the exploration of outer space as a new arena for human activity. Whether that vision persists in the twenty-first century is yet to be seen; I hope it does.

Special thanks to my editor at Penguin, Sam Raim, who knew of *Exploring the Unknown* and approached me with the idea of a one-volume "best of" the collection. Sam has been a source of encouragement and guidance throughout, and deserves my thanks for both conceiving the project and seeing it through to completion. At George Washington University, it was student assistant Liana Sherman and my colleague Allyson Reneau who helped me prepare the documents for publication. Thanks as well to Roger Launius, whose essays in *Exploring the Unknown* I have drawn on, and occasionally quoted directly, for the chapter introductions and editorial notes in this volume.

JOHN LOGSDON

Suggestions for Further Reading

NASA from its first days has had an active history program that has produced a wide variety of publications, most of which are available online in electronic form. To view the lengthy publication list, go to https://history.nasa.gov/publications.html.

All seven volumes of *Exploring the Unknown* were published by the Government Printing Office, the last in 2008. Electronic copies of each volume are online at https://history.nasa.gov/series95.html.

I have in recent years published two books dealing with presidential decisions that have shaped the U.S. space program. They are:

John M. Logsdon, *John F. Kennedy and the Race to the Moon* (Palgrave Macmillan, 2010)
John M. Logsdon, *After Apollo?: Richard Nixon and the American Space Program* (Palgrave Macmillan, 2015)

I am currently working on a third study, *Ronald Reagan and the Space Frontier*, also to be published by Palgrave Macmillan, probably in early 2019.

Many U.S. astronauts have written or cowritten books on their experiences. The best among them remains Michael Collins, *Carrying the Fire: An Astronaut's Journeys* (Farrar, Straus and Giroux, 2009). Also notable is Gene Cernan, *The Last Man on the Moon: Astronaut Gene Cernan and America's Race in Space* (St. Martin's Griffin, 2000). The most irreverent book by a shuttle astronaut is Mike Mullane, *Riding Rockets: The Outrageous Tales of a Space Shuttle Astronaut* (Scribner, 2007).

Other excellent studies of the history of the U.S. space program include:

William Burrows, *This New Ocean: The Story of the First Space Age* (Modern Library, 1999)

Andrew Chaikin, *A Man on the Moon: The Voyages of the Apollo Astronauts* (Penguin Books, 2007)

James Hansen, *First Man: The Life of Neil A. Armstrong* (Simon & Schuster, 2012)

Dennis Jenkins, *The Space Shuttle: Developing an Icon, 1972–2013* (Specialty Press, 2017)

Christopher Kraft, *Flight: My Life in Mission Control* (Dutton, 2001)

Gene Kranz, *Failure Is Not an Option: Mission Control from Mercury to Apollo 13 and Beyond* (Simon & Schuster, 2009)

W. Henry Lambright, *Powering Apollo: James E. Webb of NASA* (Johns Hopkins University Press, 1998)

Roger Launius and Howard McCurdy, *Spaceflight and the Myth of Presidential Leadership* (University of Illinois Press, 1997)

Howard McCurdy, *Space and the American Imagination* (Johns Hopkins University Press, 2011)

Howard McCurdy, *The Space Station Decision: Incremental Politics and Technological Choice* (Johns Hopkins University Press, 2007)

Walter McDougall, *. . . the Heavens and the Earth: A Political History of the Space Age* (Johns Hopkins University Press, 1997)

Yanek Mieczkowski, *Eisenhower's Sputnik Moment: The Race for Space and World Prestige* (Cornell University Press, 2013)

Charles Murray and Catherine Bly Cox, *Apollo, the Race to the Moon* (Simon & Schuster, 1989)

Michael Neufeld, *Von Braun: Dreamer of Space, Engineer of War* (Vintage, 2008)

Lynn Sherr, *Sally Ride: America's First Woman in Space* (Simon & Schuster, 2015)

Margot Lee Shetterly, *Hidden Figures: The American Dream and the Untold Story of the Black Women Mathematicians Who Helped Win the Space Race* (William Morrow, 2016)

Margaret Weitekamp, *Right Stuff, Wrong Sex: America's First Women in Space Program* (Johns Hopkins University Press, 2005)

Thomas Wolfe, *The Right Stuff* (Picador, 2008)

The Penguin Book
of Outer Space Exploration

PROLOGUE

THE DREAM OF CENTURIES[1]

Though humans have dreamed of voyaging to the stars for thousands of years, this story—in which they finally became a spacefaring species—begins in the early twentieth century, when the concept of using rocket propulsion to accelerate spacecraft to orbital velocity came to the fore. Three pioneers, Konstantin Tsiolkovsky in tsarist Russia, Hermann Oberth in post–World War I Germany, and Robert Goddard in the United States, from early in the century on, carried out theoretical work on rocketry; of the three, only Goddard translated theory into experimentation, launching the first liquid-propelled rocket in 1926. Other rocket engineering work was carried out by groups of enthusiasts, using private resources to fund their efforts. One of these groups was the German rocket society *Verein für Raumschiffahrt* (Society for Spaceship Travel, or VfR), which launched a successful rocket flight in February 1931. One of the early members of the VfR was the young Prussian nobleman Wernher von Braun. In the Soviet Union, the government-sponsored Group for the Study of Reactive Motion (GIRD), created in 1931, was the initial focal point for rocket research, launching its first liquid-fueled vehicle in 1933. One of GIRD's founders was Sergei Korolev, who would go on to be the "chief designer" of the Soviet space program. In the United States, the American Interplanetary Society (later called the American Rocket Society), another enthusiast group, carried out rocket research during the 1930s, focusing on ground-based testing of rocket engines. At the California Institute of Technology, the Guggenheim Aeronautical Laboratory (GALCIT) began work on rocket engine development, with funding eventually coming from the War Department.

It was the outbreak of World War II that altered the course of rocket development in the United States and Europe. In the

United States, GALCIT was renamed the Jet Propulsion Laboratory in 1943; its work during the war focused on using small rocket motors to assist airplane takeoffs. In the Soviet Union, from 1938 to 1944 Korolev was either imprisoned or held in gulags (slave labor camps); there was little progress in rocket research during the war. The German government after World War I had been forbidden by the Treaty of Versailles to work on developing military weapons; that prohibition did not extend to rockets, not yet considered tools of war. In 1932 the German army hired the twenty-year-old von Braun to work on military rockets. When Adolf Hitler took over the German government in 1933, von Braun continued to work for the German military. Eventually he led a team that developed the first ballistic missile, the Vengeance Weapon 2 (V-2) rocket. The V-2 was first test launched in October 1942, and from September 1944 until the end of the war in May 1945 more than three thousand V-2 rockets were launched against targets in Europe and England. Even as he worked for the Nazi regime, von Braun insisted that his primary interest was developing the rockets that would someday enable space travel.[2]

As the end of World War II approached, von Braun and key members of his rocket team, seeking to escape capture by invading Russian forces and preferring to surrender to the American military, made their way across war-torn Germany from the Baltic seacoast, where their secret base was located at Peenemunde, to Bavaria. Their surrender took place on May 3, 1945; by June, arrangements were under way to transfer von Braun and many of his associates to the United States, where they would work for the U.S. Army on rocket research. They were based near El Paso, Texas from 1946 to 1950, when the team was transferred to the Army's Redstone Arsenal in Huntsville, Alabama, home of the Army Ballistic Missile Agency. There, von Braun, rehabilitated from his collaboration with the German military and Nazi regime, became head of the agency's development operations division, and thus in charge of its rocket research.

Von Braun's work at Huntsville was just one of the postwar U.S. efforts to develop a missile capable of launching a nuclear warhead to a distant target. In the 1950s, both the Navy and the newly created Air Force sponsored research on intercontinental ballistic missiles (ICBMs). With the emergence of U.S.-Soviet rivalry, the so-called Cold War, the ability to deliver nuclear warheads over long distances was seen by both the United States and the Soviet Union as a key capability in their geopolitical contest. In the Soviet Union, Korolev, who had been freed in 1944 and quickly became a leader in Soviet space research, was also working on a high-priority basis to develop powerful rockets; in fact, because the Soviet nuclear warhead was significantly heavier than its U.S. counterpart, the Soviets needed a more powerful ICBM than those being developed in the United States. In both countries it would be a modified ICBM that would first carry a human into orbit.

With the development of powerful rockets for military purposes, the possibility of launching spacecraft, and soon after, people, into orbit and beyond was becoming technically feasible. In parallel, that possibility was fast becoming part of popular culture. As historian Roger Launius comments, "The dreams of Verne and Wells were combined with the pioneering rocketry of Goddard and Oberth and later developments in technology to create the probability of a dawning space age." In the United States, it was Wernher von Braun, just a few years removed from his service to Hitler's regime, who emerged as a leading spokesman for the future of space travel. Launius writes that von Braun's "background as a serious rocket engineer, a German émigré, a handsome aristocrat, and a charismatic leader" made him an "effective promoter of spaceflight to the public."

In October 1951 the Hayden Planetarium of the American Museum of Natural History in New York City hosted what was billed as "the First Annual Symposium on Space Travel." Among those in attendance was Gordon Manning, editor of the weekly newsmagazine *Collier's,* which together with similar newsweeklies *Life, The Saturday Evening Post*, and *Look* was a major source of information for the general public. Manning and his associates were so impressed by what they heard that they decided to run a series of articles on space travel in their magazine. The series would be edited by journalist Cornelius Ryan and would draw upon top U.S. thinkers on the potentials of space. The eight articles in the series appeared between March 22, 1952, and April 30, 1954. They were dramatically illustrated by several artists, most notably Chesley Bonestell. The initial articles in the series, reflecting the Cold War environment of the early 1950s, were cast in the context of U.S. competition with the Soviet Union for control of space.

The following two articles, one written by the magazine's editors to introduce the series and the other by Wernher von Braun, were among the first widely circulated discussions of space exploration. Von Braun's article, discussing an initial human mission to Mars some one hundred years in the future, is very speculative. The Collier's *articles, combined with a Walt Disney–produced television series based in part on them, by the mid-1950s made von Braun the most well-known spokesman on space issues and created a public expectation that human space travel would soon become reality.*

In their alarmist introduction to the series, the editors of Collier's *stressed that it was the urgency of national security and global leadership competition with Cold War rival the Soviet Union, rather than a visionary perspective on the future in space, which made it important to "conquer space." Although the visionary aspects of space activity are mentioned, it was clearly the relationship between space capability and national power that had higher priority in the view of the magazine's editors. Space travel may have been a dream of centuries, but it was rivalry on Earth that made it possible.*

Editors, "What Are We Waiting For?," *Collier's*, March 22, 1952

On the following pages *Collier's* presents what may be one of the most important scientific symposiums ever published by a national magazine. It is the story of the inevitability of man's conquest of space.

What you will read is not science fiction. It is serious fact. Moreover, it is an urgent warning that the U.S. must immediately embark on a long-range development program to secure for the West "space superiority." If we do not, somebody else will. That somebody else very probably would be the Soviet Union.

The scientists of the Soviet Union, like those of the U.S., have reached the conclusion that it is now possible to establish an artificial satellite or "space station" in which man can live and work far beyond the earth's atmosphere. In the past it has been correctly said that the first nation to do this will control the earth. And it is too much to assume that Moscow's military planners have overlooked the military potentialities of such an instrument.

A ruthless foe established on a space station could actually subjugate the peoples of the world. Sweeping around the earth in a fixed orbit, like a second moon, this man-made island in the heavens could be used as a platform from which to launch guided missiles. Armed with atomic war heads, radar-controlled projectiles could be aimed at any target on the earth's surface with devastating accuracy.

Furthermore, because of the enormous speeds and relatively small size, it would be almost impossible to intercept them. In other words: whoever is the first to build a station in space can prevent any other nation from doing likewise.

We know that the Soviet Union, like the U.S., has an extensive guided missile and rocket program under way. Recently, however the Soviets intimated that they were investigating the development of huge rockets capable of leaving the earth's atmosphere. One of their top scientists, Dr. M. K. Tikhonravov, a member of the Red Army's Military Academy of Artillery, let it be known that on the basis of Soviet scientific development such rocket ships could be built and, also, that the creation of a space station was not only feasible but definitely probable. Soviet engineers could even now, he declared, calculate precisely the characteristics of such space vehicles; and he added that Soviet developments in this field equaled, if not exceeded, those of the Western World.

We have already learned, to our sorrow, that Soviet scientists and engineers should never be underestimated. They produced the atomic bomb years earlier than was anticipated. Our air superiority over the Korean battlefields is being challenged by their excellent MIG-15 jet fighters, which, at certain altitudes, have proved much faster than ours. And while it is not believed that the Soviet Union has actually begun work on a major project to capture space superiority, U.S. scientists point out that the basic knowledge for such a program has been available for the last 20 years.

What is the U.S. doing, if anything, in this field?

In December 1948, the late James Forrestal, then Secretary of Defense, spoke of the existence of an "earth satellite vehicle program." But in the opinion of competent military observers this was little more than a preliminary study. And so far as is known today, little further progress has been made. *Collier's* feels justified in asking: What are we waiting for?

We have the scientists and the engineers. We enjoy industrial superiority. We have the inventive genius. Why, therefore, have we not embarked on a major space program equivalent to that which was undertaken in developing the atomic bomb? The issue is virtually the same.

The atomic bomb has enabled the U.S. to buy time since the end of World War II. Speaking in Boston in 1949, Winston Churchill put it this way: "Europe would have been communized and London under bombardment sometime ago but for the deterrent of the atomic bomb in the hands of the United States." The same could be said for a space station. In the hands of the West a space station, permanently established beyond the atmosphere, would be the greatest hope for peace the world has ever known. No nation could undertake preparations for war without the certain knowledge that it was being observed by the ever-watching eyes aboard the "sentinel in space." It would be the end of Iron Curtains wherever they might be.

Furthermore, the establishment of a space station would mean the dawning of a new era for mankind. For the first time, exploration of the heavens would be possible, and the great secrets of the universe would be revealed.

When the atomic bomb program—the Manhattan Project—was initiated, nobody really knew whether such a weapon could actually be made. The famous Smyth Report on atomic energy tells us that among the scientists were many who had grave and fundamental doubts of the success of the undertaking. It was a two-billion-dollar technical gamble.

Such would not be the case with a space program. The claim that huge rocket shops can be built and a space station created still stands unchallenged by any serious scientist. Our engineers can spell out right now (as you will see) the technical specifications for the rocket ship and space station in cut-and-dried figures. And they detail the design features. All they need is time (about 10 years), money and authority.

Even the cost has been estimated: $4,000,000,000. And when one considers that we have spent nearly $54,000,000,000 on rearmament since the Korean War began, the expenditure of $4,000,000,000 to produce an instrument which would guarantee the peace of the world seems negligible.

Wernher von Braun's three essays for the Collier's *series included a concept for a wheel-shaped space station in orbit more than one thousand miles above the Earth, a description of a journey to the Moon, and last, a plan to get to Mars. While von Braun was confident that humans could journey to the Moon, he was well aware of the many unsolved obstacles to successful voyages to Mars. He thus placed his initial Mars mission one hundred years in the future, sometime in the mid-twenty-first century. Von Braun's 1954 conception of a Mars mission called for a seventy-man crew. (There is no indication he conceived of a mixed-gender crew.) That crew would need a flotilla of ten spacecraft to make the Mars journey. Von Braun was sure that journey would happen . . . "someday."*

Wernher von Braun with Cornelius Ryan, "Can We Get to Mars?," *Collier's*, April 30, 1954

The first men who set out for Mars had better make sure they leave everything at home in apple-pie order. They won't get back to earth for more than two and a half years. The difficulties of a trip to Mars are formidable. The outbound journey, following a huge arc 255,000,000 miles long, will take eight months—even with rocket ships that travel many thousands of miles an hour. For more than a year, the explorers will have to live on the great red planet, waiting for it to swing into a favorable position for the return trip. Another eight months will pass before the 70 members of the pioneer expedition set foot on earth again. All during that time, they will be exposed to a multitude of dangers and strains, some of them impossible to foresee on the basis of today's knowledge.

Will man ever go to Mars? I am sure he will—but it will be a century or more before he's ready. In that time scientists and engineers will learn more about the physical and mental rigors of interplanetary flight—and about the known dangers of life on another planet. Some of that information may become available within the next 25 years or so, through the erection of a space station above the earth (where telescope viewings will not be blurred by the earth's atmosphere) and through the subsequent exploration of the moon, as described in previous issues of *Collier's*.

Even now science can detail the technical requirements of a Mars expedition down to the last ton of fuel. Our knowledge of the laws governing the solar system—so accurate that astronomers can predict an eclipse of the sun to within a fraction of a second—enables scientists to determine exactly the speed a space ship must have to reach Mars, the course that will intercept the planet's orbit at exactly the right moment, the methods to be used for the landing, take-off and other maneuvering. We know, from these calculations, that we already have chemical rocket fuels adequate for the trip.

Better propellants are almost certain to emerge during the next 100 years. In fact, scientific advances will undoubtedly make obsolete many of the engineering concepts on which this article . . . [is] based. Nevertheless, it's possible to discuss the problems of a flight to Mars in terms of what is known today. We can assume, for example, that such an expedition will involve about 70 scientists and crew members. A force that size would require a flotilla of 10 massive space ships, each weighing more than 4,000 tons—not only because there's safety in numbers, but because of the tons of fuel, scientific equipment, rations, oxygen, water and the like necessary for the trip and for a stay of about 31 months away from earth.

All that information can be computed scientifically. But science can't apply a slide rule to man; he's the unknown quantity, the weak spot that makes a Mars expedition a project for the far distant, rather than the immediate, future. The 70 explorers will endure hazards and stresses the like of which no men before them have ever known. Some of these hardships must be eased—or at least better understood—before the long voyage becomes practical.

For months at a time, during the actual period of travel, the expedition members will be weightless. Can the human body stand prolonged weightlessness? The crews of rocket ships plying between the ground and earth's space station about 1,000 miles away will soon grow accustomed to the absence of gravity—but they will experience this odd sensation for no more than a few hours at a time. Prolonged weightlessness will be a different story.

Over a period of months in outer space, muscles accustomed to fighting the pull of gravity could shrink from disuse—just as do the muscles of people who are bedridden or encased in plaster casts for a long time. The members of a Mars expedition might be seriously handicapped by such a disability. Faced with a rigorous work schedule on the unexplored planet, they will have to be strong and fit upon arrival.

The problem will have to be solved aboard the space vehicles. Some sort of elaborate spring exercisers may be the answer. Or perhaps synthetic gravity could be produced aboard the rocket ships by designing them to rotate as they coast through space, creating enough centrifugal force to act as a substitute for gravity.

Far worse than the risk of atrophied muscles is the hazard of cosmic rays. An overdose of these deep-penetrating atomic particles, which act like the invisible radiation of an atomic-bomb burst, can cause blindness, cell damage and possibly cancer.

Scientists have measured the intensity of cosmic radiation close to the earth. They have learned that the rays dissipate harmlessly in our atmosphere. They also have deduced that man can safely venture as far as the moon without risking an overdose of radiation. But that's a comparatively brief trip. What will happen to men who are exposed to rays for months on end? There is no material that offers practical protection against cosmic rays that is practical for space travel. Space engineers could provide a barrier by making the cabin walls of lead several feet thick—but that would add hundreds of tons to the weight of the space vehicle. A more realistic plan might be to surround the cabin with the fuel tanks, thus providing the added safeguard of a two- or three-foot thickness of liquid.

The best bet would seem to be a reliance on man's ingenuity; by the time an expedition from the earth is ready to take off for Mars, perhaps in the mid-2000s, it is quite likely that researchers will have perfected a drug which will enable men to endure radiation for comparatively long periods. Unmanned rockets, equipped with instruments which send information back to earth, probably will blaze the first trail to our sister planet, helping to clear up many mysteries of the journey.

* * *

Science ultimately will solve the problems posed by cosmic rays, meteors and other natural phenomena of space. But man will still face one great hazard: himself.

Man must breathe. He must guard himself against a great variety of illnesses and ailments. He must be entertained. And he must be protected from many psychological hazards, some of them still obscure.

How will science provide a synthetic atmosphere within the space-ship cabins and Martian dwellings for two and a half years? When men are locked into a confined, airtight area for only a few days or weeks oxygen can be replenished, and exhaled carbon dioxide and other impurities extracted, without difficulty. Submarine engineers solved the problem long ago. But a conventional submarine surfaces after a brief submersion and blows out its stale air. High-altitude pressurized aircraft have mechanisms which automatically introduce fresh air and expel contaminated air.

There's no breathable air in space or on Mars; the men who visit the red planet will have to carry with them enough oxygen to last many months.

WHEN MEN LIVE TOO CLOSE TOGETHER

During that time they will live, work and perform all bodily functions within the cramped confines of a rocket-ship cabin or a pressurized—and probably mobile—Martian dwelling. (I believe the first men to visit Mars will take along inflatable, spherical cabins, perhaps 30 feet across, which can be mounted atop tractor chassis.) Even with plenty of oxygen, the atmosphere in those living quarters is sure to pose a problem.

Within the small cabins, the expedition members will wash, perform personal functions, sweat, cough, cook, create garbage. Every one of those activities will feed poisons into the synthetic air—just as they do within the earth's atmosphere.

No less than 29 toxic agents are generated during the daily routine of the average American household. Some of them are body wastes, others come from cooking. When you fry an egg, the burned fat releases a potent irritant called acrolein. Its effect is negligible on earth because the amount is so small that it's almost instantly dissipated in the air. But that microscopic quantity of acrolein in the personnel quarters of a Mars expedition could prove dangerous; unless there was some way to remove it from the atmosphere it would be circulated again and again through the air-conditioning system.

Besides the poisons resulting from cooking and the like, the engineering equipment—lubricants, hydraulic fluids, plastics, the metals in the vehicles—will give off vapors which could contaminate the atmosphere.

What can be done about this problem? No one has all the answers right now, but there's little doubt that by using chemical filters, and by cooling and washing the air as it passes through the air-conditioning apparatus, the synthetic atmosphere can be made safe to live in.

Besides removing the impurities from the man-made air, it may be necessary to add a few. Man has lived so long with the impurities in the earth's atmosphere that no one knows whether he can exist without them. By the time of the Mars expedition, the scientists may decide to add traces of dust, smoke and oil to the synthetic air—and possibly iodine and salt as well.

I am convinced that we have, or will acquire, the basic knowledge to solve all the physical problems of a flight to Mars. But how about the psychological problem? Can a man retain his sanity while cooped up with many other men in a crowded area, perhaps twice the length of your living room, for more than thirty months?

Share a small room with a dozen people completely cut off from the outside world. In a few weeks the irritations begin to pile up. At the end of a few months, particularly if the occupants of the room are chosen haphazardly, someone is likely to go berserk. Little mannerisms—the way a man cracks his knuckles, blows his nose, the way he grins, talks or gestures—create tension and hatred which could lead to murder.

Imagine yourself in a space ship millions of miles from earth. You see the same people every day. The earth, with all it means to you, is just another bright star in the heavens; you aren't sure you'll ever get back to it. Every noise about the rocket ship suggests a breakdown, every crash a meteor collision. If somebody does crack, you can't call off the expedition and return to earth. You'll have to take him with you.

The psychological problem probably will be at its worst during the two eight-month travel periods. On Mars, there will be plenty to do, plenty to see. To be sure, there will be certain problems on the planet, too. There will be considerable confinement. The scenery is likely to be grindingly monotonous. The threat of danger from some unknown source will hang over the explorers constantly. So will the knowledge that an extremely complicated process, subject to possible breakdown, will be required to get them started on their way back home. Still, Columbus's crew at sea faced much the same problems the explorers will face on Mars: the fifteenth-century sailors felt the psychological tension, but no one went mad.

But Columbus traveled only ten weeks to reach America; certainly his men would never have stood an eight-month voyage. The travelers to Mars will have to, and psychologists undoubtedly will make careful plans to keep up the morale of the voyagers.

The fleet will be in constant radio communication with the earth (there probably will be no television transmission, owing to the great distance). Radio programs will help relieve the boredom, but it's possible that the broadcasts will be censored before transmission; there's no way of telling how a man might react, say, to the news that his home town was the center of a flood disaster. Knowing would do him no good—and it might cause him to crack.

Besides radio broadcasts, each ship will be able to receive (and send) radio pictures. There also will be films which can be circulated among the space ships. Reading matter will probably be carried in the form of microfilms to save space. These activities—plus frequent intership visiting, lectures and crew rotations—will help to relieve the monotony. There is another possibility, seemingly fantastic but worth mentioning briefly because experimentation already has indicated it may be practical. The nonworking members of a Mars expedition may actually hibernate during part of the long voyage. French doctors have induced a kind of artificial hibernation in certain patients for short periods in connection with operations for which they will need all their strength . . . The process involves a lowering of the body temperature, and the subsequent slowing down of all normal physical processes. On a Mars expedition, such a procedure, over a longer period, would solve much of the psychological problem, would cut sharply into the amount of food required for the trip, and would, if successful, leave the expedition members in superb physical condition for the ordeal of exploring the planet.

Certainly if a Mars expedition were planned for the next 10 or 15 years, no one would seriously consider hibernation as a solution for any of the problems of the trip. But we're talking of a voyage to be made 100 years from now; I believe that if the French experiments bear fruit, hibernation may actually be considered at that time.

Finally, there has been one engineering development which may also simplify both the psychological and physical problems of a Mars voyage. Scientists are on the track of a new fuel, useful only in the vacuum of space, which would be so economical that it would make possible far greater speeds for space journeys. It could be used to shorten the travel time, or to lighten the load of each space ship, or both. Obviously, a four- or six-month Mars flight would create far fewer psychological hazards than a trip lasting eight months. In any case, it seems certain that members of an expedition to Mars will have to be selected with great care. Scientists estimate that only one person in every 6,000 will be qualified, physically, mentally and emotionally, for routine space flight. But can 70 men be found who will have those qualities— and also the scientific background necessary to explore Mars? I'm sure of it.

One day a century or so from now, a fleet of rocket ships will take off for Mars. The trip could be made with 10 ships launched from an orbit 1,000 miles out in space that girdles our globe at its equator. (It would take tremendous power and vast quantities of fuel to leave directly from the earth. Launching a Mars voyage from an orbit about 1,000 miles out, far from the earth's gravitational pull, will require relatively little fuel.) The Mars-bound vehicles, assembled in the orbit, will look like bulky bundles of girders, with propellant tanks hung on the outside and great passenger cabins perched on top. Three of them will have torpedo-shaped noses and massive wings—dismantled, but strapped to their sides for future use. Those bullet noses will be detached and will serve as landing craft, the only vehicles that will actually land on the neighbor planet. When the 10 ships are 5,700 miles from the earth, they will cut off their rocket motors; from there on, they will coast unpowered toward Mars.

After eight months they will swing into an orbit around Mars, about 600 miles up, and adjust speed to keep from hurtling into space again. The expedition will take this intermediate step, instead of preceding directly to Mars, for two main reasons: first, the ships (except for the three detachable torpedo-shaped noses) will lack the streamlining required for flight in the Martian atmosphere; second, it will be more economical to avoid carrying all the fuel needed for the return to earth (which now comprises the bulk of the cargo) all the way down to Mars and then back up again.

Upon reaching the 600-mile orbit—and after some exploratory probings of Mars's atmosphere with unmanned rockets—the first of the three landing craft will be assembled. The torpedo nose will be unhooked, to become the fuselage of a rocket plane. The wings and set of landing skis will be attached, and the plane launched toward the surface of Mars.

The landing of the first plane will be made on the planet's snow-covered polar cap—the only spot where there is any reasonable certainty of finding a smooth surface. Once down, the pioneer landing party will unload its tractors and supplies, inflate its balloonlike living quarters, and start on a 4,000-mile overland journey to the Martian equator, where the expedition's main base will be set up (it is the most livable part of the planet well within the area that scientists want most to investigate). At the equator, the advance party will construct a landing strip for the other two rocket planes. (The first landing craft will be abandoned at the pole.)

In all, the expedition will remain on the planet 15 months. That's a long time—but it still will be too short to learn all that science would like to know about Mars.

When, at last, Mars and the earth begin to swing toward each other in the heavens, and it's time to go back, the two ships that landed on the equator will be stripped of their wings and landing gear, set on their tails and, at the proper moment, rocketed back to the 600-mile orbit on the last leg of the return journey.

What curious information will these first explorers carry back from Mars? Nobody knows—and it's extremely doubtful that anyone now living will ever know. All that can be said with certainty today is this: the trip can be made, and will be made . . . someday.

CHAPTER 1

GETTING READY FOR SPACE EXPLORATION

While Wernher von Braun and others during the 1950s were helping to create the public expectation that space travel was just around the corner, the U.S. government was taking the initial tentative steps toward making the United States a spacefaring country. With the development of powerful ballistic missiles for launching nuclear weapons getting under way, it was just a matter of time before variants of those missiles were converted into rockets for launching objects, and, soon after, people into orbit.

The United States would not be first to space, however. American prestige took a serious blow when the Soviet Union launched the first artificial Earth satellite, *Sputnik 1*, on October 4, 1957. The event was seen by many, particularly in the media and the Congress, as proof of Soviet superiority in science, technology, engineering, and social organization—with all the implications that had for the growth of Soviet power and prestige. But *Sputnik 1* did not cause high levels of alarm within the Eisenhower administration. Dwight D. Eisenhower and his associates had a different space priority. To Eisenhower and many of his associates, who had felt the impact of Pearl Harbor, minimizing the risk of a surprise Soviet attack was an overriding concern. The need to see into the Soviet Union and to learn the location and number of its military bases, nuclear facilities, missile facilities, bomber aircraft, and so forth created the demand for strategic reconnaissance on a continental scale. At that time, aircraft overflight and sending camera-carrying balloons over Soviet territory (the latter of which provided little useful intelligence) were the only means available for strategic reconnaissance, and they were forbidden by international law.

The Eisenhower administration grappled with this problem and took a number of different approaches to solve it, including a high-altitude spy plane, the U-2, thought impervious to antiaircraft attack. Another approach was approving the early development of satellites that could overfly the Soviet Union and return useful information. In 1954 the Air Force began preliminary development of a reconnaissance satellite program, dubbed WS (Weapon System) 117L. This was the first government-approved space program. Although its feasibility was far from certain at the time, it was understood that a satellite could potentially overcome the risks faced by aerial reconnaissance. First, there was no risk of a satellite getting shot down or destroyed in orbit by conventional air defenses. Second, it was not clear under international law if one country's satellite overflying another country's territory in outer space was legal or not. There was no precedent. Establishing the principle of "freedom of space," that is, the acceptability and legality of satellite overflight of another's territory, became a crucial national security issue for the United States.

Meanwhile, as these national defense concerns were being discussed inside the government, the U.S. scientific community independently was developing a proposal for a scientific satellite to be launched during the International Geophysical Year (IGY), which was to run from June 1957 to December 1958. The IGY was a major international effort to study the entire Earth, including its lands, seas, atmosphere, and outer space environments. The international scientific community would collaborate on research and share the results. Sixty-seven countries were to participate, including both the United States and the Soviet Union. To U.S. scientists, orbiting a scientific satellite was a natural extension of their post–World War II research in the upper atmosphere using high-altitude balloons and sounding rockets.

The National Science Foundation approved the National Academy of Sciences' proposal for a scientific satellite. But because of its implications for national security, the satellite program also required approval at the highest levels of the U.S. government. Thus a National Security Council paper, NSC 5520, "Draft Statement of Policy on U.S. Scientific Satellite Program," May 20, 1955, outlined a variety of reasons for approval of the scientific satellite program, including its scientific and technological benefits, its importance for national prestige, and, most important, its use in establishing the international legal precedent of freedom of space. The satellite's peaceful purposes would be emphasized, since it was to be an unclassified program and the scientific data it acquired would be shared internationally under the sponsorship of the IGY. These factors would increase the chances that no country would protest the overflight of such a satellite. With the legal precedent set, an opening for reconnaissance satellites would be created. Valid scientific interests would provide a convenient screen for equally valid national security concerns. Science would shape the environment for the military to peer behind the Iron Curtain and strengthen U.S. national security. The intentional intertwining of U.S. scientific and national security interests in the U.S. civilian space program was thus part of the U.S. space effort from its inception.

National Security Council, NSC 5520, "Draft Statement of Policy on U.S. Scientific Satellite Program," May 20, 1955

GENERAL CONSIDERATIONS

1. The U.S. is believed to have the technical capability to establish successfully a small scientific satellite of the earth in the fairly near future. Recent studies by the Department of Defense have indicated that a small scientific satellite weighing 5 to 10 pounds can be launched into an orbit about the earth using adaptations of existing rocket components. If a decision to embark on such a program is made promptly, the U.S. will probably be able to establish and track such a satellite within the period 1957–58.

2. The report of the Technological Capabilities Panel of the President's Science Advisory Committee recommended that intelligence applications warrant an immediate program leading to a very small satellite in orbit around the earth, and that re-examination should be made of the principles or practices of international law with regard to "Freedom of Space" from the standpoint of recent advances in weapon technology.

3. On April 16, 1955, the Soviet Government announced that a permanent high-level, interdepartmental commission for interplanetary communications has been created in the

Astronomic Council of the USSR Academy of Sciences. A group of Russia's top scientists is now believed to be working on a satellite program. In September 1954 the Soviet Academy of Sciences announced the establishment of the Tsiolkovsky Gold Medal which would be awarded every three years for outstanding work in the field of interplanetary communications.

4. Some substantial benefits may be derived from establishing small scientific satellites. By careful observation and the analysis of actual orbital decay patterns, much information will be gained about air drag at extreme altitudes and about the fine details of the shape of and the gravitational field of the earth. Such satellites promise to provide direct and continuous determination of the total ion content of the ionosphere. These significant findings will find ready application in defense communication and missile research. When large instrumented satellites are established, a number of other kinds of scientific data may be acquired . . .

5. From a military standpoint, the Joint Chiefs have stated their belief that intelligence applications strongly warrant the construction of a large surveillance satellite. While a small scientific satellite cannot carry surveillance equipment and therefore will not have any direct intelligence potential, it does represent a technological step toward the achievement of a large surveillance satellite, and will be helpful to this end so long as the small scientific satellite program does not impede the development of the large surveillance satellite.

6. Considerable prestige and psychological benefits will accrue to the nation which first is successful in launching a satellite. The inference of such a demonstration of advanced technology and its unmistakable relationship to intercontinental ballistic missile technology might have important repercussions on the political determination of free world countries to resist Communist threats, especially if the USSR were to be the first to establish a satellite. Furthermore, a small scientific satellite will provide a test of the principle of "Freedom of Space." The implications of this principle are being studied within the Executive Branch. However, preliminary studies indicate that

there is no obstacle under international law to the launching of such a satellite.

7. It should be emphasized that a satellite would constitute no active military offensive threat to any country over which it might pass. Although a large satellite might conceivably serve to launch a guided missile at a ground target, it will always be a poor choice for the purpose. A bomb could not be dropped from a satellite on a target below, because anything dropped from a satellite would simply continue alongside in the orbit.

8. The U.S. is actively collaborating in many scientific programs for the International Geophysical Year (IGY), July 1957 through December 1958. The U.S. National Committee of the IGY has requested U.S. Government support for the establishment of a scientific satellite during the Geophysical Year. The IGY affords an excellent opportunity to mesh a scientific satellite program with the cooperative world-wide geophysical observational program. The U.S. can simultaneously exploit its probable technological capability for launching a small scientific satellite to multiply and enhance the over-all benefits of the International Geophysical Year, to gain scientific prestige, and to benefit research and development in the fields of military weapons systems and intelligence. The U.S. should emphasize the peaceful purposes of the launching of such a satellite, although care must be taken as the project advances not to prejudice U.S. freedom of action (1) to proceed outside the IGY should difficulties arise in the IGY procedure, or (2) to continue with its military satellite programs directed toward the launch of a large surveillance satellite when feasible and desirable.

9. The Department of Defense believes that, if preliminary design studies and initial critical component development are initiated promptly, sufficient assurance of success in establishing a small scientific satellite during the IGY will be obtained before the end of this calendar year to warrant a response, perhaps qualified, to an IGY request. The satellite itself and much information as to its orbit would be public information. The means of launching would be classified.

10. A program for a small scientific satellite could be developed from existing missile programs already underway within the Department of Defense. Funds of the order of $20 million are estimated to be required to give reasonable assurance that a small scientific satellite can be established during 1957–8 . . .

COURSES OF ACTION

11. Initiate a program in the Department of Defense to develop the capability of launching a small scientific satellite by 1958, with the understanding that this program will not prejudice continued research toward large instrumented satellites for additional research or intelligence capabilities, or materially delay other major Defense programs.

12. Endeavor to launch a small scientific satellite under international auspices, such as the International Geophysical Year, in order to emphasize its peaceful purposes, provided such international auspices are arranged in a manner which:
    a. Preserves U.S. freedom of action in the field of satellites and related programs.
    b. Does not delay or otherwise impede the U.S. satellite program and related research and development programs.
    c. Protects the security of U.S. classified information regarding such matters as the means of launching a scientific satellite.
    d. Does not involve actions which imply a requirement for prior consent by any nation over which the satellite might pass in its orbit, and thereby does not jeopardize the concept of "Freedom of Space."

This statement was approved at a May 26, 1955, meeting of the National Security Council, thereby becoming in effect the first U.S. National Space Policy.

A proposal from the civilian-like Naval Research Laboratory (NRL) for a small scientific satellite developed by the scientific community, funded by the National Science Foundation, and launched on top of a modified sounding rocket, Viking, was selected on September 9, 1955, as the U.S. space contribution to the IGY. The satellite project's name was Vanguard, and the first launch was expected to be in late 1957 or early 1958.

Vanguard was selected in the face of a competing proposal. The potential existed for the U.S. Army, using a rocket developed by Wernher von Braun and his German rocket team, to launch a satellite in late 1956 or early 1957. The Army Ballistic Missile Agency's (ABMA) proposal was called Project Orbiter. But the Eisenhower administration judged that a U.S. military satellite, launched on a military ballistic missile, before the beginning of the IGY, and without scientific justification, was likely to be viewed as a provocative act by the Soviets, generate a Soviet protest, and thus fail to establish the freedom of space principle. There was also some consideration that it should not be von Braun and his team, so recently employed by Nazi Germany, who should be responsible for this U.S. accomplishment. President Eisenhower decided that the establishment of the freedom of space principle was more important than being first. The Army protested the decision to choose the Navy's Project Vanguard, but to no avail.

Such considerations did not constrain Soviet behavior. Sergei Korolev, the "chief designer" of the Soviet space program, and his associates were determined to be the first into space. On October 4, 1957, the Soviet Union launched the first Earth-orbiting satellite. This "simple satellite," as it was called by Korolev, was explicitly built to beat the United States to the first satellite launch. The Soviets called the satellite "Sputnik," or "fellow traveler," and reported the achievement in a tersely worded press release issued by the official news agency, Tass, printed in the October 5, 1957, issue of the Communist Party newspaper Pravda. *The press release ended by linking the satellite launch to the success of the "new socialist society."*

"Announcement of the First Satellite," *Pravda,* October 5, 1957

For several years scientific research and experimental design work have been conducted in the Soviet Union on the creation of artificial satellites of the earth. As already reported in the press, the first launching of the satellites in the USSR were planned for realization in accordance with the scientific research program of the International Geophysical Year. As a result of very intensive work by scientific research institutes and design bureaus the first artificial satellite in the world has been created. On October 4, 1957, this first satellite was successfully launched in the USSR. According to preliminary data, the carrier rocket has imparted to the satellite the required orbital velocity of about 8000 meters per second. At the present time the satellite is describing elliptical trajectories around the earth, and its flight can be observed in the rays of the rising and setting sun with the aid of very simple optical instruments (binoculars, telescopes, etc.). According to calculations which now are being supplemented by direct observations, the satellite will travel at altitudes up to 900 kilometers above the surface of the earth; the time for a complete revolution of the satellite will be one hour and thirty-five minutes; the angle of inclination of its orbit to the equatorial plane is 65 degrees. On October 5 the satellite will pass over the Moscow area twice—at 1:46 a.m. and at 6:42 a.m. Moscow time. Reports about the subsequent movement of the first artificial satellite launched in the USSR on October 4 will be issued regularly by broadcasting stations. The satellite has a spherical shape 58 centimeters in diameter and weighs 83.6 kilograms. It is equipped with two radio transmitters continuously emitting signals at frequencies of 20.005 and 40.002 megacycles per second (wave lengths of about 15 and 7.5 meters, respectively). The power of the transmitters ensures reliable reception of the signals by a broad range of radio amateurs. The signals have the form of telegraph pulses of about 0.3 second's duration with a pause of the same duration. The signal of one frequency is sent during the pause in the signal of the other frequency. Scientific stations located at various points in the Soviet Union are tracking the satellite and determining the elements of its trajectory. Since the density of the rarefied upper layers of the atmosphere is not accurately known, there are no data at present for the precise determination of the satellite's lifetime and of the point of its entry into the dense layers of the atmosphere. Calculations have shown that owing to the tremendous velocity of the satellite, at the end of its existence it will burn up on reaching the dense layers of the atmosphere at an altitude of several tens of kilometers. As early as the end of the nineteenth century the possibility of realizing cosmic flights by means of rockets was first scientifically substantiated in Russia by the works of the outstanding Russian scientist K. E. Tsiolkovskii. The successful launching of the first man-made earth satellite makes a most important contribution to the treasure-house of world science and culture. The scientific experiment accomplished at such a great height is of tremendous importance for learning the properties of cosmic space and for studying the earth as a planet of our solar system. During the International Geophysical Year the Soviet Union proposes launching several more artificial earth satellites. These subsequent satellites will be larger and heavier and they will be used to carry out programs of scientific research. Artificial earth satellites will pave the way to interplanetary travel and, apparently, our contemporaries will witness how the freed and conscientious labor of the people of the new socialist society makes the most daring dreams of mankind a reality.

The legality of overflight was suddenly no longer an issue; the United States did not protest Sputnik's overflight of American territory. The international legal precedent had been set. The way was now clear for the United States to develop its reconnaissance satellite program and thereby gain the means to observe the interior of the Soviet Union.

The public and Congress were shocked by the launch of Sputnik. America's sense of security was shattered as Sputnik visibly demonstrated that the United States could be directly threatened by a nation half a world away. The Eisenhower administration had been aware of the imminent launch of the first Soviet satellite and had given some thought to potential public reaction to such an event. But when the launch occurred on October 4, 1957, the administration was surprised by the amount of public concern about the security implications of the Soviet achievement. A crisis atmosphere erupted in the media and in Congress, and although Eisenhower did not agree that the launch of Sputnik 1 warranted an urgent response, he was forced politically into reacting with a series of rapid steps.

Four days after the event, Secretary of State John Foster Dulles sent White House press secretary James Hagerty his suggestions for the text of a press release that would place the Sputnik launch in a nonthreatening context and reassure the public. Although the core of the German "rocket team," including Wernher von Braun, had surrendered to the United States at the end of World

War II, Dulles attributes the Soviet success to German engineers working in the Soviet Union. Although Dulles's comments did not result in a press release, they did form the basis for much of the administration's "official" reaction to the Soviet achievement as well as the core of Eisenhower's comments at a press conference on October 9, after the president had met with his advisers to discuss the significance of Sputnik.

John Foster Dulles, "Draft Statements on the Soviet Satellite," October 8, 1957

The launching by the Soviet Union of the first earth satellite is an event of considerable technical and scientific importance. However, that importance should not be exaggerated. What has happened involves no basic discovery and the value of a satellite to mankind will for a long time be highly problematical. That the Soviet Union was first in this project is due to the high priority which the Soviet Union gives to scientific training and to the fact that since 1945 the Soviet Union has particularly emphasized developments in the fields of missiles and of outer space. The Germans had made a major advance in this field and the results of their effort were largely taken over by the Russians when they took over the German assets, human and material, at Peenemünde, the principal German base for research and experiment in the use of outer space. This encouraged the Soviets to concentrate upon developments in this field with a use of resources and effort not possible in time of peace to societies where the people are free to engage in pursuits of their own choosing and where public monies are limited by representatives of the people. Despotic societies which can command the activities and resources of all their people can often produce spectacular accomplishments. These, however, do not prove that freedom is not the best way. While the United States has not given the same priority to outer space developments as has the Soviet Union, it has not neglected this field. It already has a capability to utilize outer space for missiles and it is expected to launch an earth satellite during the present geophysical year in accordance with a program which has been under orderly development over the past two years. The United States welcomes the peaceful achievement of the Soviet scientists. It hopes that the acclaim which has resulted from their effort will encourage the Soviet Union to seek development along peaceful lines and seek to enrich the spiritual and material welfare of their people. What is happening with reference to outer space makes more than ever important the proposal made by the United States and the other free world members of the Disarmament Subcommittee. I recall my White House statement of August 28 which emphasized the proposal of the Western Powers at London to establish a study group to the end that "outer space shall be used only for peaceful, not military, purposes."

General Andrew Goodpaster, Eisenhower's military aide, took notes during the first discussion between President Eisenhower and his top national security advisers after the Soviet launch of Sputnik 1. Deputy Secretary of Defense Donald Quarles, who among the group was the most well versed on space matters, led the discussion. Eisenhower indicated that he had never worried about being first to space, but he also recognized that the United States had to respond to the Soviet challenge.

Andrew Goodpaster, "Memorandum of Conference with the President," October 8, 1957

Secretary Quarles began by reviewing a memorandum prepared in Defense for the President on the subject of the earth satellite (dated October 7, 1957). He left a copy with the President. He reported that the Soviet launching on October 4th had apparently been highly successful.

The President asked Secretary Quarles about the report that had come to his attention to the effect that Redstone could have been used and could have placed a satellite in orbit many months ago. Secretary Quarles said there was no doubt that the Redstone, had it been used, could have orbited a satellite a year or more ago. The Science Advisory Committee had felt, however, that it was better to have the earth satellite proceed separately from military development. One reason was to stress the peaceful character of the effort, and a second was to avoid the inclusion of materiel, to which foreign scientists might be given access, which is used in our own military rockets. He said that the Army feels it could erect a satellite four months from now if given the order—this would still be one month prior to the estimated date for the Vanguard. The President said that when this information reaches the Congress, they are bound to ask why this action was not taken. He recalled, however, that timing was never given too much importance in our own program, which was tied to the IGY and confirmed that in order for all scientists to be able to look at the instrument, it had to be kept away from military secrets. Secretary Quarles pointed out that the Army plan would require some modification of the instrumentation in the missile.

He went on to add that the Russians may have in fact done us a good turn, unintentionally, in establishing the concept of freedom of international space—this seems to be generally accepted as orbital space, in which the missile is making an inoffensive passage.

The President asked what kind of information could be conveyed by the signals reaching us from the Russian satellite. Secretary Quarles said the Soviets say that it is simply a pulse to permit location of the missile through radar direction finders . . .

The President asked the group to look ahead five years, and asked about a reconnaissance vehicle. Secretary Quarles said the Air Force has a research program in this area and gave a general description of the project.

Governor Adams recalled that Dr. Pusey had said that we had never thought of this as a crash program, as the Russians apparently did. We were working simply to develop and transmit scientific knowledge. The President thought that to make a sudden shift in our approach now would be to belie the attitude we have had all along. Secretary Quarles said that such a shift would create service tensions in the Pentagon. Mr. Holaday said he planned to study with the Army the back up of the Navy program with the Redstone, adapting it to the instrumentation.

There was some discussion concerning the Soviet request, as to whether we would like to put instruments of ours aboard one of their satellites . . . Several present pointed out that our instruments contain parts which, if made available to the Soviets, would give them substantial technological information.

As a first step in responding to the Soviets, the United States needed to get a satellite into orbit. The first Vanguard launch attempt on December 6, 1957, ended in a humiliating failure in front of the TV cameras. The Army, having never given up on its ambition to launch a satellite into orbit, had after Sputnik been authorized to attempt a satellite launch, and on January 31, 1958, successfully used von Braun's Jupiter-C rocket (a modified Redstone intermediate range ballistic missile) to put Explorer 1 *into orbit.*

By February 1958, four months after Sputnik 1, *how best to organize for the U.S. response to Sputnik was still uncertain. It was likely that a new space agency would be created; however, its responsibilities, form, and organizational location were still undecided. The question of the military or civilian character of a new agency was discussed in a meeting among the president, vice president, other White House officials, and Republican leaders in Congress. Eisenhower's newly appointed science adviser, James Killian, and Vice President Richard Nixon were advocates of a separate civilian space agency, while the president was undecided on which organizational path to follow. He was also skeptical of carrying out U.S. space projects on a crash basis to compete with Soviet achievements.*

L. A. Minnich Jr., "Legislative Leadership Meeting, Supplementary Notes," February 4, 1958

A question was raised as to whether a new Space Agency should be set up within Department of Defense (as provided in the pending Defense appropriation bill), or be set up as an independent agency. The President's feeling was essentially a desire to avoid duplication, and priority for the present would seem to rest with Defense because of paramountcy of defense aspects. However, the President thought that in regard to non-military aspects, Defense could be the operational agent, taking orders from some non-military scientific group. The National Science Foundation, for instance, should not be restricted in any way in its peaceful research.

Dr. Killian had some reservations as to the relative interest and activity of military vs. peaceful aspects, as did the Vice President who thought our posture before the world would be better if non-military research in outer space were carried forward by an agency entirely separate from the military.

There was some discussion of the prospect of a lunar probe. Dr. Killian thought this might be next on the list of Russian efforts. He had some doubt as to whether the United States should at this late date attempt to press a lunar probe, but the question would be fully canvassed by the Science Advisory Committee in the broad survey it had under way. Dr. Killian thought the United States might do a lunar probe in 1960, or perhaps get to it on a crash program by 1959. Sen. Saltonstall had heard, however, that it might even be accomplished in 1958, if pressed hard enough.

**

The President was firmly of the opinion that a rule of reason had to be applied to these Space projects—that we couldn't pour unlimited funds into these costly projects where there was nothing of early value to the Nation's security. He recalled the great effort he had made for the Atomic Peace Ship but Congress would not authorize it, even though in his opinion it would have been a very worthwhile project.

And in the present situation, the President mused, he would rather have a good Redstone [a missile carrying a nuclear warhead] than be able to hit the moon, for we didn't have any enemies on the moon.

Sen. Knowland pressed the question of hurrying along with a lunar probe, because of the psychological factor. He recalled the great impact of Sputnik, which seemed to negate the impact of our large mutual security program. If we are close enough to doing a probe, he said, we should press it. The President thought it might be OK to go ahead with it if it could be accomplished with some missile already developed or nearly ready, but he didn't want to just rush into an all-out effort on each one of these possible glamor performances without a full appreciation of their great cost. Also, there would have to be a clear determination of what agency would have the responsibility.

The Vice President reverted to the idea of setting up a separate agency for "peaceful" research projects, for the military would be deterred from things that had no military value in sight. The President thought Defense would inevitably be involved since it presently had all the hardware, and he did not want further duplication. He did not preclude having eventually a great Department of Space.

The Eisenhower administration also began organizing the U.S. government for the space age. A first action, taken soon after the Soviet Union on November 3, 1957, launched the large Sputnik 2 carrying the dog Laika, was to create both a White House position of presidential science adviser and also a part-time advisory group of eminent scientists called the President's Science Advisory Committee (PSAC). One of PSAC's initial assignments was to examine the rationales for space activity and recommend how to organize the United States for space. After much deliberation and consideration of various alternatives, three separate space activity paths were developed. One track included military space activities. On February 7, 1958, these activities were assigned to a newly created Advanced Research Projects Agency (ARPA) within the Department of Defense. Eisenhower hoped that establishing this new agency with authority over all military space efforts would prevent intense interservice rivalry over the space mission. Another track approved the same day was a top secret strategic reconnaissance program to observe the Soviet Union, called CORONA; this program was based on one element of WS-117L, a photo intelligence satellite that would drop exposed film from orbit, where it would be recovered in midair. Management of CORONA was assigned to a covert CIA and Air Force team similar to that managing the U-2 spy aircraft program; this team subsequently would become the core of the secret National Reconnaissance Office.

The third track, chosen after a careful examination of alternatives, was to create a separate civilian space program, focused on science and exploration activities, conducted on an unclassified basis, and open to international cooperation, with a new civilian space agency to manage it. It is worth noting that based upon the debates and decisions made in the immediate aftermath of Sputnik, the U.S. government space program was divided into the civilian, military, and intelligence space sectors that still exist today, more than sixty years later.

There are many reasons why the United States decided to create a separate civilian space agency. Initially, President Eisenhower reasoned that the primary focus of the space program would be military and intelligence missions, and that a separate civilian agency might not be necessary. But he was soon convinced that a civilian space program was important both for the scientific aspects of space and for domestic and international audiences. The United States could demonstrate how open, scientific, and peaceful its purposes in space were, as opposed to the closed, militaristic nature of the Soviet program. Furthermore, the U.S. scientific community did not trust the U.S. military to have its best interests in mind when it made its budgetary and priority decisions. Eisenhower agreed. Besides, the potential for international cooperation in space science would be unduly constrained in a scientific program under military control. Finally, putting a civilian face on American space activities would help divert attention from Eisenhower's primary goal for the space program: getting strategic intelligence on the Soviet Union.

With the requirement for a civilian space agency decided, the next question was how to organize it. With the military option ruled out, three primary alternatives were considered. The first option was to create a new government agency entirely from scratch. But it would take too much time to legislate, fund, organize, and build new facilities. The second possibility was to assign civilian space activities to an expanded Atomic Energy Commission (AEC). But the AEC had virtually no experience with space-related programs, and civilian space activities would interfere with its current, vital national security function. It was finally agreed that expansion of the National Advisory Committee for Aeronautics (NACA) offered the best solution. Although it did not have extensive rocket or space experience and capabilities, NACA had a number of factors in its favor. NACA was not part of the Department of Defense. It had been working on flight technology since its inception during World War I, it had a history of successfully managing interactions with the military and industry, it had an extensive network of research installations across the United States, and it had developed a preliminary proposal for a comprehensive, well-reasoned space program. By early March 1958, draft legislation was in the works for the creation of a space agency built upon the NACA structure.

The following memorandum was the culmination of discussion on how best to organize the nation's space effort, and it laid the basis for President Eisenhower's decision to create a new civilian space agency based on NACA.

James R. Killian Jr. (Special Assistant for Science and Technology), Percival Brundage (Director, Bureau of the Budget), Nelson A. Rockefeller (Chairman, President's Advisory Committee on Government Organization), Memorandum for the President, "Organization for Civil Space Programs," March 5, 1958, with attached: "Summary of Advantages and Disadvantages of Alternative Organizational Arrangements"

THE PROBLEM

As you know, there will soon be presented for your consideration civil space programs for the United States which will entail increased expenditures and the employment of important numbers of scientists, engineers and technicians.

This Committee, in conjunction with the Director of the Bureau of the Budget and your Special Assistant for Science and Technology, have given consideration to the manner in which the executive branch should be organized to conduct the new program. This memorandum contains our joint findings and recommendations. The memorandum (1) discusses some of the factors which should be taken into account in establishing the government's organization for these civil space programs, (2) recommends a pattern of organization, and (3) indicates certain interim actions which will be necessary. Also attached is a summary of the advantages and disadvantages of certain alternative organizational arrangements.

Discussions to date suggest that an aggressive space program will produce important civilian gains in the form of advances in general scientific knowledge and protection of the international prestige of the United States. These benefits will be in addition to such military uses of outer space as may prove feasible.

ESTABLISHING A LONG TERM ORGANIZATION

Because of the importance of the civil interest in space exploration, the long term organization for Federal programs in this area should be under civilian control. Such civilian domination is also suggested by public and foreign relations considerations. However, civilian control does not envisage taking out from military control projects relating to missiles, anti-missile defense, reconnaissance satellites, military communications, and other space technology relating to weapons systems or direct military requirements.

We have considered a number of different approaches to civil space organization. It is our conclusion that one of these alternatives provides a workable solution to the problem. The other principal alternatives have serious shortcomings which argue against their selection as a basis for space organization.

Recommendation No. l. We recommend that leadership of the civil space effort be lodged in a strengthened and re-designated National Advisory Committee for Aeronautics.

The National Advisory Committee for Aeronautics (NACA), in a resolution adopted on January 16, 1958, has proposed that the national space program be implemented by the cooperative effort of the Department of Defense, the NACA, the National Academy of Sciences and the National Science Foundation, together with the universities, research institutions, and industrial companies of the nation. NACA further recommended that the development of space vehicles and the operations required for scientific research in space phenomena and space technology be conducted by the NACA when within its capabilities. NACA is now formulating a program which is expected to propose expansion of existing programs and the addition of supplementary research facilities.

FACTORS FAVORING NACA AS THE PRINCIPAL CIVIL SPACE AGENCY

1. NACA is a going Federal research agency with a large scientific and engineering staff (approximately 2,000 of its 7,500 employees are in these categories) and a large plant ($300,000,000 in laboratories and test facilities). It can expand its research program and increase its emphasis on space matters with a minimum of delay and can provide a functioning institutional setting for this activity.

2. NACA's aeronautical research has been progressively involving it in technical problems associated with space flight and its current facilities construction program is designed to be useful in space research. It has done research in rocket engines (including advanced chemical propellants), it has

developed materials and designs to withstand the thermal effects of high speeds in or on entering the earth's atmosphere, it conducts multi-stage rocket launchings, and in the X-15 project it has taken the leadership (in cooperation with the Navy and Air Force) in developing a manned vehicle capable of flights beyond the earth's atmosphere.

3. If NACA is not given the leading responsibility for the civil space program, its future research role will be limited to aircraft and missiles. Some of its present activities would have to be curtailed, and the logical paths of progress in much of its current work would be closed. It would, under such circumstances, be difficult for NACA to attract and retain the most imaginative and competent scientific and engineering personnel, and all aspects of its mission could suffer. Moreover, it is questionable whether it would be possible to define practicable boundaries between the missile and high performance aircraft research now performed by NACA and the space vehicle projects.

4. NACA has a long history of close and cordial cooperation with the military departments. This cooperation has taken place under a variety of arrangements, usually with little in the way of formalized agreements. Although new relationship problems are bound to arise from an augmented NACA role in space programs, the tradition of comity and civil-military accommodation which has been built up over the years will be a great asset in minimizing friction between the civil space agency and the Department of Defense.

5. Although much of its work has been done for the military departments, NACA is a civilian agency and widely recognized as such. A civilian setting for space programs is desirable and NACA satisfies this requirement.

***

SUMMARY OF ADVANTAGES AND DISADVANTAGES OF ALTERNATIVE ORGANIZATIONAL ARRANGEMENTS

1. Use of a private contractor to carry out the civil space program under supervision of NACA.

A variation of our recommended organizational approach is to select NACA as the civilian agency to supervise contracts with a private laboratory charged with developing and testing space vehicles. This is the pattern followed by the Atomic Energy Commission in much of its research. This approach has also been used to some extent by the military services in developing missiles.

Advantages

Contract operation is preferred by some scientific personnel as a means of circumventing government salary and administrative controls. It would retain NACA in a supervisory capacity while making use of selected private research organizations.

Disadvantages

This approach is in conflict with the traditional NACA practice of carrying out research largely through its own government-employee staffed laboratories: there is no assurance that a private research laboratory can be found to do the work on a sufficiently urgent schedule; and such greater flexibility as private laboratories may enjoy can also be provided NACA through the changes in law previously described.

Conclusion

No real gains would flow from this alternative which could not be achieved under the preferred organization. It would be better to permit NACA to make its own decisions as to the extent to which it would use contracting authority in executing the space research program. It is assumed, of course, that NACA will, in fact, make fairly extensive use of research contracts, but on a selective basis.

2. Utilization of the Department of Defense

The recent Supplemental Military Construction Authorization Act authorizes the Secretary of Defense, for a period of one year, to carry on such space projects as may be designated by the President. It confers permanent authority for the Secretary or his designee to proceed with missile and other space projects directly related to weapons systems and military requirements.

Advantages

The Department of Defense is now doing most of the current missile and satellite work: it has the bulk of the scientists and engineers active in these fields in its employ or on the rolls of its contractors; it will have to continue work on space vehicles on an interim basis for demonstration purposes; it is experienced in working with and utilizing the facilities of NACA; and it may be possible for a civilian agency of the Department to carry out the program.

Disadvantages

The Department of Defense is a military agency in law and in the eyes of the world and placing the space program under it would be interpreted as emphasizing military goals: the space program is expected to produce benefits largely unrelated to the central mission of the Department of Defense; there is some danger that the non-military phases of space activity would be neglected; the Department is already so overloaded with its central military responsibilities that care should be taken to avoid charging it with additional civil functions; cooperation with other nations in international civil space matters could be made more difficult; and adequate civil-military cooperation can be achieved under the recommended organization without assigning inappropriate functions to Defense.

Conclusion

Since the space program has a relatively limited military significance, at least for the foreseeable future, and since the general scientific objectives should not be subordinated to military priorities, it is essential that the arrangements for space organization provide for leadership by a civilian agency.

3. Utilization of the Atomic Energy Commission

There are now pending before the Congress bills which would authorize the Atomic Energy Commission to proceed with the development of vehicles for the exploration of outer space. Among these bills are S.3117 (introduced by Senator Anderson) and S.3000 (introduced by Senator Gore). The justification for these proposals is the role already being played by the Atomic Energy Commission in developing nuclear propelled jet and rocket engines.

Advantages
The Atomic Energy Commission is a civilian agency with competence in directing scientific research and development projects: it has had experience in managing research contracts and in working with the military agencies; and it is now charged with developing a nuclear rocket engine which may eventually be used to propel space vehicles.

Disadvantages
The Atomic Energy Commission is concerned chiefly with the use of a single form of energy and it is expected that chemical propellants, not atomic energy, will be the chief power source for space vehicles for years to come. Moreover, the Commission has virtually no experience or competence in most aspects of the design, construction and testing of space vehicles.

Conclusion
The Atomic Energy Commission has a contribution to make in the space field. However, it should limit its work to the aspects of the space problem in which nuclear energy may have practical applications. An administration position along these lines has already been conveyed to the Chairman of the Atomic Energy Commission.

4. Creation of a Department of Science and Technology

Senators Humphrey, McClellan and Yarborough recently introduced S.3126, a bill to create a Department of Science and Technology. The bill calls for the establishment of a new executive department which at the outset would contain or be given the functions of the National Science Foundation, the Patent

Office, the Office of Technical Services of the Department of Commerce, the National Bureau of Standards, the Atomic Energy Commission and certain divisions of the Smithsonian Institution. The Secretary would also be authorized to establish institutes for basic research.

Advantages

The proposed department would provide a civilian setting for the administration of space programs, and it would give this and other scientific activities the prestige and accessibility to the President associated with departmental status.

Disadvantages

The proposed department will be highly controversial, and there is no assurance that it can be established in time to assume the responsibility for civil space programs. It is also unlikely that science, of itself, will provide a sound basis for organizing an executive department.

Conclusion

There would be little prospect of getting such a reorganization approved and functioning in the near future. Even if the department could be created, it might not provide as good a setting for a high priority space program as that proposed under the preferred organization.

In parallel with these organizational steps, the Eisenhower administration considered the initial programs the new space agency would undertake and placed them in a comprehensive policy framework. The President's Science Advisory Committee assessed the appropriate direction and pace for the U.S. space program. Not surprisingly, given that its members were prominent scientists, the committee focused heavily on the scientific aspects of the space program. With the president's endorsement, on March 26, 1958, PSAC released its report outlining the importance of space activities. The committee recommended a measured pace in pursuing a U.S. space effort, wanting to prevent the diversion of resources to the space program from other areas of scientific activity.

President's Science Advisory Committee, "Introduction to Outer Space," March 26, 1958

What are the principal reasons for undertaking a national space program? What can we expect to gain from space science and exploration? What are the scientific laws and facts and the technological means which it would be helpful to know and understand in reaching sound policy decisions for a United States space program and its management by the Federal Government? This statement seeks to provide brief and introductory answers to these questions.

It is useful to distinguish among four factors which give importance, urgency, and inevitability to the advancement of space technology. The first of these factors is the compelling urge of man to explore and to discover, the thrust of curiosity that leads men to try to go where no one has gone before. Most of the surface of the earth has now been explored and men now turn to the exploration of outer space as their next objective.

Second, there is the defense objective for the development of space technology. We wish to be sure that space is not used to endanger our security. If space is to be used for military purposes, we must be prepared to use space to defend ourselves.

Third, there is the factor of national prestige. To be strong and bold in space technology will enhance the prestige of the United States among the peoples of the world and create added confidence in our scientific, technological, industrial, and military strength.

Fourth, space technology affords new opportunities for scientific observation and experiment which will add to our knowledge and understanding of the earth, the solar system, and the universe.

The determination of what our space program should be must take into consideration all four of these objectives. While this statement deals mainly with the use of space for scientific inquiry, we fully recognize the importance of the other three objectives.

In fact it has been the military quest for ultra-long-range rockets that has provided man with new machinery so powerful that it can readily put satellites in orbit and, before long, send instruments out to explore the moon and nearby planets. In this way, what was at first a purely military enterprise has opened up an exciting era of exploration that few men, even a decade ago, dreamed would come in this century.

***

WILL THE RESULTS JUSTIFY THE COSTS?

Since the rocket power plants for space exploration are already in existence or being developed for military need, the cost of additional scientific research, using these rockets, need not be exorbitant. Still, the cost will not be small, either. This raises an important question that scientists and the general public (who will pay the bill) both must face: Since there are still so many unanswered scientific questions and problems all around us on earth, why should we start asking new questions and seeking out new problems in space? How can the results possibly justify the cost?

Scientific research, of course, has never been amenable to rigorous cost accounting in advance. Nor, for that matter, has exploration of any sort. But if we have learned one lesson, it is that research and exploration have a remarkable way of paying off—quite apart from the fact that they demonstrate that man is alive and insatiably curious. And we all feel richer for knowing what explorers and scientists have learned about the universe in which we live. It is in these terms that we must measure the value of launching satellites and sending rockets into space.

***

The scientific opportunities are so numerous and so inviting that scientists from many countries will certainly want to participate. Perhaps the International Geophysical Year will suggest a model for the international exploration of space in the years and decades to come.

The timetable suggests the approximate order in which some of the scientific and technical objectives mentioned in this review may be attained.

The timetable is not broken down into years, since there is yet too much uncertainty about the scale of the effort that will be made. The timetable simply lists various types of space investigations and goals under three broad headings: Early, Later, Still Later . . .

EARLY
1. Physics
2. Geophysics
3. Meteorology
4. Minimal Moon Contact
5. Experimental Communications
6. Space Physiology

LATER
1. Astronomy
2. Extensive Communications
3. Biology
4. Scientific Lunar Investigation
5. Minimal Planetary Contact
6. Human Flight in Orbit

STILL LATER
1. Automated Lunar Exploration
2. Automated Planetary Exploration
3. Human Lunar Exploration and Return

AND MUCH LATER STILL
Human Planetary Exploration

In conclusion, we venture two observations. Research in outer space affords new opportunities in science, but it does not diminish the importance of science on earth. Many of the secrets of the universe will be fathomed in laboratories on earth, and the progress of our science and technology and the welfare of the Nation require that our regular scientific programs go forward without loss of pace, in fact at an increased pace. It would not be in the national interest to exploit space science at the cost of weakening our efforts in other scientific endeavors. This need not happen if we plan our national program for space science and technology as part of a balanced national effort in all science and technology.

Our second observation is prompted by technical considerations. For the present, the rocketry and other equipment used in space technology must usually be employed at the very limit of its capacity. This means that failures of equipment and uncertainties of schedule are to be expected. It therefore appears wise to be cautious and modest in our predictions and pronouncements about future space activities—and quietly bold in our execution.

The Eisenhower administration used its National Security Council mechanism to prepare statements of national policy in various areas of government activity. These statements were usually classified and intended to reflect administration internal thinking and to provide policy guidance to government agencies. A series of statements of National Space Policy were developed in the aftermath of Sputnik 1. These statements were in general more bullish on the potential impacts of space activity than were the personal views of President Eisenhower, who remained skeptical of the need for a fast-paced U.S. space program intended to compete in a "space race" with the Soviet Union.

The first policy statement appeared in June 1958, even as the Congress was still considering President Eisenhower's April 1958 proposal to create what became NASA. With its recognition of the political and security impacts of Soviet leadership in space, this statement coming from the national security community is very different in tone from the public statements of the Eisenhower administration, which was trying to minimize the significance of Soviet space achievements. The statement is also notable for its early discussion of the nontechnical impacts of space exploration.

National Security Council, NSC 5814, "U.S. Policy on Outer Space," June 20, 1958

INTRODUCTORY NOTE

This Statement of U.S. Policy on Outer Space is designated Preliminary because man's understanding of the full implications of outer space is only in its preliminary stages. As man develops a fuller understanding of the new dimension of outer space, it is probable that the long-term results of exploration and exploitation will basically affect international and national political and social institutions.

Perhaps the starkest facts which confront the United States in the immediate and foreseeable future are (1) the USSR has surpassed the United States and the Free World in scientific and technological accomplishments in outer space, which have captured the imagination and admiration of the world; (2) the USSR, if it maintains its present superiority in the exploitation of outer space, will be able to use that superiority as a means of undermining the prestige and leadership of the United States; and (3) the USSR, if it should be the first to achieve a significantly superior military capability in outer space, could create an imbalance of power in favor of the Sino-Soviet Bloc and pose a direct military threat to U.S. security.

The security of the United States requires that we meet these challenges with resourcefulness and vigor.

GENERAL CONSIDERATIONS
INTRODUCTION
Significance of Outer Space to U.S. Security

1. More than by any other imaginative concept, the mind of man is aroused by the thought of exploring the mysteries of outer space.
2. Through such exploration, man hopes to broaden his horizons, add to his knowledge, and improve his way of living on earth. Already, man is sure that through further exploration he can obtain certain scientific and military values. It is reasonable for man to believe that there must be, beyond these areas, different and great values still to be discovered.
3. The technical ability to explore outer space has deep psychological implications over and above the stimulation provided by the opportunity to explore the unknown. With its hint of the possibility of the discovery of fundamental truths concerning man, the earth, the solar system, and the universe, space exploration has an appeal to deep insights within man which transcend his earthbound concerns. The manner in which outer space is explored and the uses to which it is put thus take on an unusual and peculiar significance.
4. The beginning stages of man's conquest of space have been focused on technology and have been characterized by national competition. The result has been a tendency to equate

achievement in outer space with leadership in science, military capability, industrial technology, and with leadership in general.

5. The initial and subsequent successes by the USSR in launching large earth satellites have profoundly affected the belief of peoples, both in the United States and abroad, in the superiority of U.S. leadership in science and military capability. This psychological reaction of sophisticated and unsophisticated peoples everywhere affects U.S. relations with its allies, with the Communist Bloc, and with neutral and uncommitted nations.

6. In this situation of national competition and initial successes by the USSR, further demonstrations by the USSR of continuing leadership in outer space capabilities might, in the absence of comparable U.S. achievements in this field, dangerously impair the confidence of these peoples in U.S. overall leadership. To be strong and bold in space technology will enhance the prestige of the United States among the peoples of the world and create added confidence in U.S. scientific, technological, industrial and military strength.

7. The novel nature of space exploitation offers opportunities for international cooperation in its peaceful aspects. It is likely that certain nations may be willing to enter into cooperative arrangements with the United States. The willingness of the Soviets to cooperate remains to be determined. The fact that the results of cooperation in certain fields, even though entered into for peaceful purposes, could have military application, may condition the extent of such cooperation in those fields.

***

Manned Exploration of Outer Space

24. In addition to satisfying man's urge to explore new regions, manned exploration of outer space is of importance to our national security because:

   a. Although present studies in outer space can be carried on satisfactorily by using only unmanned vehicles, the time will undoubtedly come when man's judgment and resourcefulness will be required fully to exploit the potentialities of outer space.

b. To the layman, manned exploration will represent the true conquest of outer space. No unmanned experiment can substitute for manned exploration in its psychological effect on the peoples of the world.

c. Discovery and exploration may be required to establish a foundation for the rejection of USSR claims to exclusive sovereignty of other planets which may be visited by nationals of the USSR.

25. The first step in manned outer space travel could be undertaken using rockets and components now under study and development. Travel by man to the moon and beyond will probably require the development of new basic vehicles and equipment.

The Eisenhower administration sent its proposed National Aeronautics and Space Act of 1958 to Congress on April 2. The bill was passed in modified form and signed by Eisenhower on July 29. At the insistence of Senate majority leader Lyndon B. Johnson, the Space Act created a National Aeronautics and Space Council, chaired by the president, to coordinate U.S. civil and national security space activities. The Space Act was amended in following years, but its statement of the goals of U.S. space activity has withstood the test of time.

National Aeronautics and Space Act of 1958, July 29, 1958

To provide for research into problems of flight within and outside the earth's atmosphere, and for other purposes.

Be it enacted by the Senate and House of Representatives of the United States of America in Congress assembled,

TITLE I—SHORT TITLE, DECLARATION OF POLICY, AND DEFINITIONS

SHORT TITLE

Sec. 101. This act may be cited as the "National Aeronautics and Space Act of 1958."

DECLARATION OF POLICY AND PURPOSE

Sec. 102. (a) The Congress hereby declares that it is the policy of the United States that activities in space should be devoted to peaceful purposes for the benefit of all mankind.

b. The Congress declares that the general welfare and security of the United States require that adequate provision be made for aeronautical and space activities. The Congress further declares that such activities shall be the responsibility of, and shall be directed by, a civilian agency exercising control over aeronautical and space activities sponsored by the United States, except that activities peculiar to or primarily associated with the development of weapons systems, military operations, or the defense of the United States (including the research and development necessary to make effective provision for the defense of the United States) shall be the responsibility of, and shall be directed by, the Department of Defense; and that determination as to which such agency has responsibility for and direction of any such activity shall be made by the President in conformity with section 201 (e).

c. The aeronautical and space activities of the United States shall be conducted so as to contribute materially to one or more of the following objectives:

1. The expansion of human knowledge of phenomena in the atmosphere and space;

2. The improvement of the usefulness, performance, speed, safety, and efficiency of aeronautical and space vehicles;

3. The development and operation of vehicles capable of carrying instruments, equipment, supplies and living organisms through space;

4. The establishment of long-range studies of the potential benefits to be gained from, the opportunities for, and the problems involved in the utilization of aeronautical and space activities for peaceful and scientific purposes.

5. The preservation of the role of the United States as a leader in aeronautical and space science and technology and in the application thereof to the conduct of peaceful activities within and outside the atmosphere.

6. The making available to agencies directly concerned with national defenses of discoveries that have military value or significance, and the furnishing by such agencies, to the civilian agency established to direct and control nonmilitary aeronautical and space activities, of information as to discoveries which have value or significance to that agency;

7. Cooperation by the United States with other nations and groups of nations in work done pursuant to this Act and in the peaceful application of the results, thereof; and

8. The most effective utilization of the scientific and engineering resources of the United States, with close cooperation among all interested agencies of the United States in order to avoid unnecessary duplication of effort, facilities, and equipment.

***

TITLE II—COORDINATION OF AERONAUTICAL AND SPACE ACTIVITIES

NATIONAL AERONAUTICS AND SPACE COUNCIL

Sec. 201. (a) There is hereby established the National Aeronautics and Space Council (hereinafter called the "Council") which shall be composed of—

1. the President (who shall preside over meetings of the Council);

2. the Secretary of State;

3. the Secretary of Defense

4. the Administrator of the National Aeronautics and Space Administration;

5. the Chairman of the Atomic Energy Commission;

6. not more than one additional member appointed by the President from the departments and agencies of the Federal Government; and

7. not more than three other members appointed by the President, solely on the basis of established records of distinguished achievement from among individuals in private life

who are eminent in science, engineering, technology, education, administration, or public affairs.

***

NATIONAL AERONAUTICS AND SPACE ADMINISTRATION

Sec. 202. (a) There is hereby established the National Aeronautics and Space Administration (hereinafter called the "Administration"). The Administration shall be headed by an Administrator, who shall be appointed from civilian life by the President by and with the advice and consent of the Senate, and shall receive compensation at the rate of $22,500 per annum. Under the supervision and direction of the President, the Administrator shall be responsible for the exercise of all powers and the discharge of all duties of the Administration, and shall have authority and control over all personnel and activities thereof.

***

On October 1, 1958, the new NASA opened its doors for business. The new space agency took over most of the civilian projects that had been under ARPA's interim management since earlier in the year. NASA also took control of existing NACA infrastructure, including Langley Aeronautical Lab in Hampton, Virginia; Ames Aeronautical Lab in Mountain View, California; Wallops Island rocket range on the Delmarva Peninsula; the Lewis Flight Propulsion Lab in Cleveland, Ohio; and the High-Speed Flight Propulsion station at Edwards Air Force Base, California. Funds were provided for the construction of a new spaceflight center in Greenbelt, Maryland (soon named after Robert Goddard), intended to be the operations center for most NASA missions. The staff of the Vanguard program was transferred from the Naval Research Laboratory to Goddard as the initial core of its staff. The human spaceflight mission, after some controversy, was transferred to NASA from the Air Force. NASA was also provided with the state-of-the-art capabilities and personnel the Army had at the Jet Propulsion Laboratory (JPL) of the California Institute of Technology. JPL was a premier engineering organization that had developed the first U.S.

satellite. By the end of 1959, over the objections of the Army leadership, von Braun's German rocket team and other personnel at the Army Ballistic Missile Agency in Huntsville, Alabama, were also transferred to NASA management, becoming the George C. Marshall Space Flight Center.

T. Keith Glennan, president of the Case Institute of Technology in Cleveland, was selected as NASA's first administrator. Glennan was an engineer who prior to World War II had worked primarily in the motion picture industry. He had one previous tour of duty in Washington, as a member of the Atomic Energy Commission from 1950 to 1952. In this excerpt from the diary he prepared for his daughters, Glennan discusses how he came to NASA, his initial impressions of the tasks before him, and the philosophy he brought to the position.

T. Keith Glennan, excerpts from *The Birth of NASA: The Diary of T. Keith Glennan*

Imagine my surprise when on 7 August 1958 I received a call from Jim Killian asking me to come immediately to Washington. I flew down on that same day and met with him at his apartment that evening. He said his purpose was to ask me, on behalf of President Eisenhower, to consider becoming administrator of the new agency, which of course was the National Aeronautics and Space Administration (NASA). He handed me a copy of the bill, which I had not previously seen. I read it through rather hurriedly and pointed out immediately the built-in conflict that seemed to me to be present whereby the Defense Department most certainly would dispute the claim of the civilian agency to important elements of any program that might be initiated. After some considerable discussion, I agreed to meet with the president the next morning.

The meeting with President Eisenhower was brief and very much to the point. He said he wanted to develop a program that would be sensibly paced and vigorously prosecuted. He made no mention of concern over accomplishments of the Soviet Union although it was clear he was concerned about the nature and quality of scientific and technological progress in this country. He seemed to rely on the advice of Jim Killian. I agreed that I would give the matter consideration and would give him a reply within a few days.

***

After two or three days of soul searching I called Killian to say I would accept—but only if Hugh Dryden, the director of the NACA, would endorse the appointment and would agree to serve as my deputy. Events began to move rapidly . . .

The swearing-in was set for 19 August in Washington . . . I found myself thrust into the problems of the new agency. Dryden called in Abe Silverstein and some of the top operating people who wanted to discuss budget. I will not try to describe the budget cycle in Washington agencies; suffice it to say that we were attempting to put together a budget that should have been initiated months before. Staff members were seeking my approval of a figure toward which they might work on the budget, which had to be submitted within weeks. Imagine my consternation when they proposed that we seek $615 million. The Case budget at that time was in the neighborhood of $6 or $7 million and I doubt that I had much feel for $615 million. Members of the staff made the point that when NASA was to be declared "ready for operations" we would be taking over from the Defense Department projects, together with manpower and funds already appropriated. It appeared that we would have about $300 million for FY 1959 (July 1958–June 1959). Their arguments must have been convincing, for I approved a budget for FY 1960 using the guideline figure of $615 million. This, then, was my introduction to what was to become one of the major activities of the federal government.

***

As I look back over my appointment schedules for those days, I wonder how I kept anything straight. I was concerned with acquiring a number of good men to fill top positions in the agency and I seem to have spent a good bit of my time on this task. Hardly a day passed without a visit from the representatives of some industrial concern—usually the president—and meetings with top people in the Department of Defense and some of the other agencies with which we would be dealing.

Although NACA contained many fine technical people, it had been an agency protected from the usual infighting found on the Washington scene. Its staff, composed of able people, had little depth and little experience in the management of large projects. Considerable thought had been given by the staff to the organization that might develop, and these plans served to get us underway. It became apparent almost immediately that further studies would be needed and that some good people would have to be hired.

Let me discuss the philosophy with which I approached this job—a philosophy about which I had thought while vacationing at Martha's Vineyard. First, having the conviction that our government operations were growing too large, I determined to avoid excessive additions to the federal payroll. Since our organizational structure was to be erected on the NACA staff, and their operation had been conducted almost wholly "in-house," I knew I would face demands on the part of our technical staff to add to in-house capacity. Indeed, approval had been given in the budget to initiate construction of a so-called "space control center" laboratory at Beltsville, Maryland, an action I approved. But I was convinced that the major portion of our funds must be spent with industry, education, and other institutions. Second, it seemed to me that we were starting virtually from scratch and with little in the way of rocket-propelled launching systems. Thus it seemed to me that we should mount an aggressive program that would build on the advancing state of the art as we came to understand more about technologies with which we were dealing. Third, it seemed clear that we should not lose sight of the propaganda values residing in successful launches—yet we had to be aware of the limitations imposed upon us by the lack of availability of proven launch vehicle systems. This was because the military missile program was just reaching the testing stage and these same rocket-propelled units were going to have to serve as "booster systems" or, as we came to call them, launch vehicle systems for our space shots. Fourth, in the nature of things it seemed necessary that we structure our program in accord with our own ideas of fields to be explored and the pace at which progress could and should be made. This meant we must avoid the undertaking of particular shots, the purpose of which would be propagandistic rather than directed toward solid accomplishment. Fifth, we faced the prospect of carrying to completion the projects started by the Advanced Research Projects Agency of the Defense Department, called into being by Secretary Neil H. McElroy during the period between 4 October 1957 and the operational beginnings of NASA. At the same time we must be planning our own broadly-based program of science and technology and organizing to accomplish all these tasks.

***

As NASA began operations, it was also developing ambitious plans for its future. Those ambitions were constrained by the overall strategy the Eisenhower administration had developed for its Cold War competition with the Soviet Union. That strategy revolved around pursuing long-term economic, military, international, and social and moral objectives that would enhance the United States' position as the world leader. It represented a commitment to putting constant pressure on the Soviet Union on a broad front, but refrained from fomenting any confrontation that might require nuclear war to resolve. A key ingredient of this strategy involved not responding to every situation vis-à-vis the Soviet Union as a crisis. Accordingly, Eisenhower had resisted the crisis sentiment that Sputnik and the early space race fostered among many policy makers in Washington. This memorandum captures the spirit of that resistance by reporting on the president's 1959 questioning of NASA's proposed budget for Fiscal Year 1961, which would begin on July 1, 1960.

A. J. Goodpaster (Brigadier General), Memorandum of Conference with the President, September 29, 1959

Others present: Dr. Kistiakowsky [Presidential Science Adviser, replacing James Killian], General Goodpaster

The President began by saying [he] had heard that Dr. Glennan is putting in for some $800 million in the FY-61 budget for space activities. He thought this was much too great an increase over the current year and in fact said that he thought a program at a rather steady rate of about a half-billion dollars a year is as much as would make sense. Dr. Kistiakowsky said that he has had about the same figure in his mind, but pointed out that this amount would not allow enough funds for space "spectaculars" to compete psychologically with the Russians, while being a great deal more than could be justified on the basis of scientific activity in relation to other scientific activities.

The President recalled that he has been stressing that we should compete in one or two carefully selected fields in our space activity, and not scatter our efforts across the board. He observed that other countries did not react to the Russian Sputnik the way the U.S. did (in fact, it was the U.S. hysteria that had most affect [sic] on other countries). Even the United States did not react very greatly to the Soviet "Lunik"—the shot that hit the moon.

The President said he had understood that, through the NASA taking over ABMA [the Army Ballistic Missile Agency, the home of Wernher von Braun's German rocket team], there was supposed to be a saving of money, but that it appeared this would in fact increase the NASA budget. He thought that Dr. Glennan should be talked to about this, away from his staff, who are pushing a wide range of projects, and advised not to overstress the psychological factor. The President thought we should take the "man in space project" and concentrate on it. He added he did not see much sense to the U.S. having more than one "super-booster" project. There should be only one.

Dr. Kistiakowsky said he strongly agreed on this. He pointed out that, by putting ABMA into NASA, there would be an over-all saving of money. The President reiterated that there is need for a serious talk with Dr. Glennan. He thought ABMA should be transferred to NASA and that we should pursue one big-booster project. Our concentration should be on real scientific endeavor. In the psychological field, we should concentrate on one project, plus the natural "tangents" thereto. He thought perhaps Dr. Glennan is overrating the need for psychological impact projects. Dr. Kistiakowsky said that the Defense Department states that if Dr. Glennan does not push fast enough in space activities, Defense will do so. The President said we must also talk to Dr. York [Herbert York was head of research and engineering in the Department of Defense], and call on him to exercise judgment. He asked that a meeting be set up, to be attended by Dr. Glennan, Dr. Dryden, Dr. Kistiakowsky, Dr. York, and Secretary Gates, in about ten days. The President stressed that we must think of the maintenance of a sound economy as well as the desirability of all these projects. He thought perhaps NASA sights are being set too high, including too many speculative projects.

While consolidation of NASA's physical infrastructure was occurring, NASA was also busy charting a course for its future. The National Aeronautics and Space Act of 1958 set out very general objectives and goals for NASA, leaving plenty of room for it to develop its own agenda. To address this situation, by December 1959 NASA produced its first long-range plan. This plan called for a program of research, development, and operations in space that would eventually make possible "the manned exploration of the moon and the nearby planets." This initial ten-year plan for NASA, developed during the agency's first year of operation, set out an ambitious program of robotic and human spaceflight, culminating in a human mission to the Moon sometime after 1970.

NASA, Office of Program Planning and Evaluation, "The Long Range Plan of the National Aeronautics and Space Administration," December 16, 1959

INTRODUCTION

The long-term national objectives of the United States in aeronautical and space activities are stated in general terms in the enabling legislation establishing NASA. It is the responsibility of NASA to interpret the legislative language in more specific terms . . .

In operational terms, these objectives are instructions to explore and to utilize both the atmosphere and the regions outside the earth's atmosphere for peaceful and scientific purposes, while at the same time providing research support to the Department of Defense. These objectives can be attained only by means of a broad and soundly conceived program of research, development and operations in space. In the long run, such activities should make feasible the manned exploration of the moon and the nearby planets, and this exploration may thus be taken as a long-term goal of NASA activities. To assure steady and rapid progress toward these objectives, a NASA Long Range Plan has been developed and is presented in this document . . .

Table 1. Nasa Mission Target Dates

Calendar Year	
1960	First launching of a Meteorological Satellite.
	First launching of a Passive Reflector Communications Satellite.
	First launching of a Scout vehicle.
	First launching of a Thor-Delta vehicle.
	First launching of an Atlas-Agena-B vehicle (by the Department of Defense).
	First suborbital flight of an astronaut.
1961	First launching of a lunar impact vehicle.
	First launching of an Atlas-Centaur vehicle.
1961–1962	Attainment of manned space flight, Project Mercury.
1962	First launching to the vicinity of Venus and/or Mars.
1963	First launching of two stage Saturn vehicle.
1963–1964	First launching of unmanned vehicle for controlled landing on the moon.
	First launching Orbital Astronomical and Radio Astronomy Observatory.
1964	First launching of unmanned lunar circumnavigation and return to earth vehicle.
	First reconnaissance of Mars and/or Venus by an unmanned vehicle.
1965–1967	First launching in a program leading to manned circumlunar flight and to permanent near-earth space station.
Beyond 1970	Manned flight to the moon.

At the same time that the NASA long-range plan was being developed, a broader discussion was occurring at the top of the U.S. government on what the goals of the civilian space program should be. Eventually, the debate boiled down to two paths that could be taken. One path was to engage the Soviet Union in a space race aimed at spectacular space achievements in order to reclaim America's prestige. The other path was to focus on scientifically justified objectives, some of which could also help re-create an image of U.S. technological leadership after the shock of early Soviet space successes. In January 1960 the Eisenhower administration issued a comprehensive statement of national space policy that tried to balance these two alternatives paths.

This final Eisenhower administration statement on U.S. space policy was adopted at a joint meeting of the National Aeronautics and Space Council and the National Security Council on January 12, 1960, and was approved by the president on January 26, 1960. It replaced earlier National Security Council policy statements. This document was the last formal statement of National Space Policy until President Jimmy Carter issued such a statement on May 11, 1978. The statement provides a comprehensive look at the state of space activities as of 1960 and was remarkably prescient with respect to the many issues that would have to be dealt with in coming years. It was originally classified "Secret" and almost completely declassified more than three decades later.

This policy gave priority to tangible payoffs rather than competition for prestige; still, it also acknowledged that it was in the U.S. interest not to allow the Soviet Union an unchallenged opportunity to exploit space for prestige purposes. By the end of 1960, NASA had the infrastructure, expertise, and programs to engage in a race for prestige with the Soviets, but not a presidential directive to do so. As will be seen in chapter 3, that was soon to change.

National Aeronautics and Space Council, "U.S. Policy on Outer Space," January 26, 1960

GENERAL CONSIDERATIONS

Scope of Policy

1. This policy is concerned with U.S. interests in scientific, civil, military, and political activities related to outer space. It deals with sounding rockets, earth satellites, and outer space vehicles, their relationship to the exploration and use of outer space, and their political and psychological significance. Although the relation between outer space technology and ballistic missile technology is recognized, U.S. policy on ballistic missiles is not covered in this policy. Anti-missile defense

systems also are not covered except to the extent that space vehicles may be used in connection with such systems.

Significance of Outer Space to U.S. Security

2. Outer space presents a new and imposing challenge. Although the full potentialities and significance of outer space remain largely to be explored, it is already clear that there are important scientific, civil, military, and political implications for the national security, including the psychological impact of outer space activities which is of broad significance to national prestige.

3. Outer space generally has been viewed as an area of intense competition which has been characterized to date by comparison of Soviet and U.S. activities. The successes of the Soviet Union in placing the first earth satellite in orbit, in launching the first space probe to reach escape velocity, in achieving the first "hard" landing on the moon and in obtaining the first pictures of the back side of the moon have resulted in substantial and enduring gains in Soviet prestige. The U.S. has launched a greater number of earth satellites and has also launched a space probe which has achieved escape velocity. These U.S. activities have resulted in a number of scientifically significant "firsts." However, the space vehicles launched by the Soviet Union have been substantially heavier than those in the U.S., and weight has been a major point of comparison internationally. In addition, the Soviets have benefited from their ability to conceal any failures from public scrutiny.

4. From the political and psychological standpoint the most significant factor of Soviet space accomplishments is that they have produced new credibility for Soviet statements and claims. Where once the Soviet Union was not generally believed, even its boldest propaganda claims are now apt to be accepted at face value, not only abroad but in the United States. The Soviets have used this credibility for the following purposes:

    a. To claim general superiority for the Soviet system on the grounds that the Sputniks and Luniks demonstrate the ability of the system to produce great results in an extremely short period of time.

    b. To claim that the world balance has shifted in favor of Communism.

    c. To claim that Communism is the wave of the future.

    d. To create a new image of the Soviet Union as a technologically powerful, scientifically sophisticated nation that is equal to the U.S. in most respects, superior in others, and with a far more brilliant future.

    e. To create a new military image of the vast manpower of the Communist nations now backed by weaponry that is as scientifically advanced as that of the West, superior in the missile field, and superior in quantity in all fields.

5. Soviet efforts already have achieved a considerable degree of success, and may be expected to show further gains with each notable space accomplishment, and particularly each major "first."

6. Significant advances have been made in restoring U.S. prestige overseas, and in increasing awareness of the scope and magnitude of the U.S. outer space effort. Although most opinion still considers the U.S. as probably leading in general scientific and technical accomplishments, the USSR is viewed in most quarters as leading in space science and technology. There is evidence that a considerable portion of world leadership and the world public expects the United States to "catch up" with the Soviet Union, and further expects this to be demonstrated by U.S. ability to equal Soviet space payloads and to match or surpass Soviet accomplishments. Failure to satisfy such expectations may give rise to the belief that the United States is "second best," thus transferring to the Soviets additional increments of prestige and credibility now enjoyed by the United States.

7. To the layman, manned space flight and exploration will represent the true conquest of outer space and hence the ultimate goal of space activities. No unmanned experiment can substitute for manned space exploration in its psychological effect on the peoples of the world. There is no reason to believe that the Soviets, after getting an earlier start, are placing as much emphasis on their manned space flight program as is the U.S.

8. The scientific value of space exploration and the prestige accruing therefrom have been demonstrated. The scientific uses of space are a potent factor in the derivation of fundamental

information of use in most fields of knowledge. Further, the greater the breadth and precision of the knowledge of the space environment, the greater the ability to exploit its potentials.

9. Among several foreseeable civil applications of earth satellites, two at present offer unique capabilities which are promising in fields of significance to the national economy: communications and meteorology. Other civil potentials are also likely to be identified.

10. The great importance of certain military utilization of outer space already has been recognized; however, the full military potential of outer space remains to be determined by further experience, studies, technical developments and strategic considerations. Space technology constitutes a foreseeable means of obtaining increasingly essential information regarding a potential enemy whose area and security preclude the effective and timely acquisition of these data by foreseeable non-space techniques. Space technology is being further utilized with the intention of more effectively accomplishing other military functions by complementing or extending non-space capabilities. In addition, as space technology and resulting uses of outer space expand, new military requirements and opportunities for development of new military capabilities are likely to materialize.

11. Space vehicles may also have important application and may play a key role in the implementation of international agreements which may be concluded respecting the reduction and control of armaments, cessation of atomic tests, and safeguards against surprise attack.

12. Outer space activities present new opportunities and problems in the conduct of the relations of the U.S. with its allies, neutral states, and the Soviet Bloc; and the establishment of sound international relationships in this new field is of fundamental significance to the national security. Of importance in seeking such relationships is the fact that all nations have an interest in the purposes for which outer space is explored and used and in the achievement of an orderly basis for the conduct of space activities. Moreover, many nations are capable of participating directly in various

aspects of outer space activities, and international participation in such applications of space vehicles as those involved in scientific research, weather forecasting, and communications may be essential to full realization of the potentialities of such activities. In addition, an improvement of the international position of the U.S. may be effected through U.S. leadership in extending internationally the benefits of the peaceful purposes of outer space. The fact that the results of the arrangements in certain fields, even though entered into for peaceful purposes, could have military implications, may condition the extent of such arrangements in those fields.

USE OF OUTER SPACE
General

13. As further knowledge of outer space is obtained, the advantages to be accrued will become more apparent. At the present time, space activities are directed toward technological development and scientific exploration; however, it is anticipated that systems will be put into operation, beginning in the near future, that will more directly contribute to national security and well-being and be of international benefit.

***

Operational Applications of Space Technology

18. All applications of the technology of outer space that now show promise of early operational utility for military or civilian purposes are based on the earth satellite. These applications ultimately will have to meet one of several criteria if they are to survive in either the defense program or the civilian economy. They will have to make possible the more efficient operation of an existing activity or create a new and desirable activity. It is expected that benefits will be gained from these applications, but the full extent of their military, economic, political and social implications has yet to be determined. Military applications are designed to

enhance military capabilities by fulfilling stated requirements of the Military Services and are currently being developed for use as operational systems. The applications
that are expected to be available earliest are as follows:

- **Meteorology**—Satellite systems to provide weather
  data on a global scale, making use of such techniques
  as television, optics, infrared detectors and radar. Information on cloud cover, storm locations, precipita-
  tion, wind direction, heat balance and water vapor
  would permit improved weather forecasting, including
  storm warnings, useful in a variety of civil activities
  such as agricultural, industrial and transportation activities, and would provide weather information to
  meet military operational needs.

- **Communications**—Satellite systems to improve and
  extend existing world-wide communications. For the
  Military Services, such systems would provide more
  effective global military communications for purposes
  of command, control, and support of military forces.
  Civil applications will benefit through more prompt
  service, increased message capacity, and greater reliability. Direct world-wide transmission of voice and
  video signals is envisaged.

- **Navigation**—Satellite systems to provide global all-
  weather capability, for land, sea and air vehicles,
  which will permit accurate determination of position;
  in the case of the military, secure operations would be
  possible . . .

Manned Space Flight and Exploration

20. It is expected that manned space flight will add significantly
    to the effectiveness of many of the scientific, military and
    civil applications indicated in the foregoing paragraphs.
    There are a number of important reasons why manned
    space activities, including the initial step of placing a man
    in orbit, are being carried out. Primary among these are:

    a. To the layman, manned space flight and exploration
       will represent the true conquest of outer space. No
       unmanned experiment can substitute for manned

exploration in its psychological effect on the peoples of the world.

b. Man's judgment, decision-making capability, and resourcefulness will ultimately be needed in many instances to ensure the full exploitation of space technology.

Moreover, manned space flight is required for scientific studies in which man himself is the principal subject of the experiment, because there is no substitute for the conduct in outer space of essential psychological and biological studies of man.

INTERNATIONAL PRINCIPLES, PROCEDURES AND ARRANGEMENTS

21. National policies and international agreements have dealt extensively with "air space" and expressly assert national sovereignty over this region; however, the upper limit of air space has not been defined. The term "outer space" also has no accepted definition, and the consequences of adopting a definition cannot now be fully anticipated. Although an avowedly arbitrary definition might prove useful for specific purposes, most of the currently foreseeable legal problems of outer space may be resolved without a precise line of demarcation between air space and outer space.

***

26. There is frequent and sharpening concern on the part of world opinion over the military implications of unchecked competition in outer space between the U.S. and the Soviet Union, and there is an accompanying interest in international agreements, controls or restrictions to limit the dangers felt to stem from such competition. With regard to the armaments control aspects of outer space, the United States first proposed in 1957, in connection with international consideration of an armaments control system, that a multilateral technical committee be set up to attempt to design an inspection system to ensure that the sending of objects through outer space will be exclusively for peaceful purposes. Furthermore, the United States has offered, if there is

a general agreement to proceed with this study without awaiting the conclusion of negotiations on other substantive disarmament proposals. There has not, to date, been multilateral agreement to proceed with such a study, and U.S. policy has not been determined concerning either the scope of control and inspection required to ensure that outer space could be used only for peaceful purposes or the relationship of any such control arrangement to other aspects of an arms agreement.

27. Exploration and use of celestial bodies require separate consideration. Neither the U.S. nor any other state has yet taken a position regarding the questions of whether a celestial body is capable of appropriation to national sovereignty and if so what acts would suffice to found a claim thereto. It is clear that serious problems would arise if a state claimed, on one ground or another, exclusive rights over all or part of a celestial body. At an appropriate time some form of international arrangement may prove useful.

28. Other problems in which all states have an interest arise from the operation of space vehicles. The following problems appear amenable to early treatment with a view to seeking internationally a basis for orderly accomplishment of space vehicle operations: (a) identification and registration of space vehicles; (b) liability for injury or damage caused by space vehicles; (c) reservation of radio frequencies for space vehicles and the related problem of termination of transmission; (d) avoidance of interference between space vehicles and aircraft; and (e) the re-entry and landing of space vehicles, through accident or design, on the territory of other states.

29. Although only a few states may be capable of mounting comprehensive outer space efforts, many states are capable of participating in the conduct of outer space activities, and active international cooperation in selected activities offers scientific, economic, and political opportunities. Continuation and extension of such cooperation in the peaceful uses of outer space through a variety of governmental and non-governmental arrangements should further enhance the position of the United States as the leading advocate of the exploration and use of outer space for the benefit of all.

Where space vehicles are employed for military applications, some degree of international cooperation may also prove useful. Any international arrangements for cooperation in outer space activities may require determination of the net advantage to U.S. security.

30. The role most appropriately undertaken by the United Nations with respect to the foregoing matters appears to lie in performing two principal functions: (a) facilitating international cooperation in the exploration and use of outer space, and (b) providing a forum for consultation and agreement respecting international problems arising from outer space activities. Future developments may make it desirable for additional functions to be performed by or under the auspices of the United Nations.

Objectives

31. Carry out energetically a program for the exploration and use of outer space by the U.S., based upon sound scientific and technological progress, designed: (a) to achieve that enhancement of scientific knowledge, military strength, economic capabilities, and political position which may be derived through the advantageous application of space technology and through appropriate international cooperation in related matters, and (b) to obtain the advantages which come from successful achievements in space.

***

Psychological Exploitation

36. To minimize the psychological advantages which the USSR has acquired as a result of space accomplishments, select from among those current or projected U.S. space activities of intrinsic military, scientific or technological value, one or more projects which offer promise of obtaining a demonstrably effective advantage over the Soviets and, so far as is consistent with solid achievements in the overall space program, stress these projects in present and future programming.

37. Identify, to the greatest extent possible, the interests and aspirations of other Free World nations in outer space with U.S.-sponsored activities and accomplishments.
38. Develop information programs that will exploit fully U.S. outer space activities on a continuing basis; especially develop programs to counter overseas the psychological impact of Soviet outer space activities and to present U.S. outer space progress in the most favorable light.

***

CHAPTER 2

FIRST STEPS [3]

In parallel with the 1958–60 steps to get NASA up and running, the new organization began its most visible effort, sending Americans into space. While getting the human spaceflight assignment was a top priority of the newly established NASA, it was not necessarily a given that it would be assigned that role. Although NASA's predecessor NACA had been examining issues related to human spaceflight during the early 1950s and had developed a plan for an initial human spaceflight program in the aftermath of *Sputnik 1*, it faced a determined rival for the lead spaceflight role in the U.S. Air Force. The Air Force had developed a proposal for a "Man-in-Space-Soonest" (MISS) effort, and fought hard within the Washington bureaucracy for getting the assignment to carry out that proposal. In February 1958, President Eisenhower had created within the Department of Defense the Advanced Research Projects Agency (ARPA) and assigned to it all U.S. space missions. In turn ARPA had assigned the human spaceflight role to the Air Force. Before it became clear that NACA would become the core of the new space agency proposed by President Eisenhower, it seemed to NACA leadership that it might be the junior partner to the Air Force in human spaceflight. But Eisenhower, as he sent his proposal for a new space agency to Congress, had said that "the new Agency will be given responsibility for all programs except those peculiar to or primarily associated with military weapons systems or military operations."

There were several months of controversy over whether human spaceflight was "primarily associated with . . . military operations." The Air Force argued that outer space was simply an extension of the atmosphere as an arena for military activities, and thus piloted activities in outer space were logically part of their mission. The service even began to use the word "aerospace" to suggest that air and space were one continuous medium. This memo, reflecting the Air Force's views, is the first formal document establishing a U.S. human spaceflight program.

General Thomas D. White (Chief of Staff, USAF), and Hugh L. Dryden (Director, NACA), Memorandum of Understanding, "Principles for the Conduct by the NACA and the Air Force of a Joint Project for a Recoverable Manned Satellite Vehicle," April 29, 1958

A. A project for a recoverable manned satellite test vehicle shall be conducted jointly by the NACA and the Air Force, implementing an ARPA instruction to the Air Force of February 23, 1958. Accomplishment of this project is a matter of national urgency.

B. The objectives of the project shall be:
   1. To achieve manned orbital flight at the earliest practicable date consistent with reasonable safety for the man,
   2. To evaluate factors affecting functions and capabilities of man in an orbiting vehicle,
   3. To determine functions best performed by man in an orbiting weapon system.

C. To insure that these objectives are achieved as early and as economically as possible the NACA and the Air Force will each contribute their specialized scientific, technical, and administrative skills, organization and facilities.

D. Overall technical direction of the project shall be the responsibility of the Director, NACA, acting with the advice and assistance of the Deputy Chief of Staff, Development, USAF.

E. Financing of the design, construction, and operational phases of the project shall be the function of the Air Force.

F. Management of the design, construction, and operational phases of the project shall be performed by the Air Force in accordance with the technical direction prescribed in paragraph D. Full use shall be made of the extensive background and capabilities of the Air Force in the Human Factors area.

G. Design and construction of the project shall be accomplished through a negotiated contract (with supplemental prime or sub-contracts) obtained after evaluating competitive proposals invited from competent industry sources. The basis for soliciting proposals will be characteristics jointly evolved by the Air Force and NACA based on studies already well under way in the Air Force and the NACA.

H. Flights with the system shall be conducted jointly by the NACA, the Air Force, and the prime contractor, with the program being directed by the NACA and the Air Force. The NACA shall have final responsibility for instrumentation and the planning of the flights.

I.   The Director, NACA, acting with the advice and assistance of the Deputy Chief of Staff, Development, USAF, shall be responsible for making periodic progress reports, calling conferences, and disseminating technical information and results of the project by other appropriate means subject to the applicable laws and executive orders for the safeguarding of classified information.

The head of Eisenhower's Bureau of the Budget, Maurice Stans, disagreed with the Air Force, pointing out that it was the president's intent to give NASA the lead in all areas not clearly related to national defense. At first, NACA, in its transition to NASA, believed it would have to play a supporting role in the spaceflight initiative, yielding the management and policy lead in human activities to the military, but it became clear as 1958 progressed that Stans was correct and that the White House wanted NASA to take the lead.

Maurice H. Stans (Director, Bureau of the Budget), Memorandum for the President, "Responsibility for 'Space' Programs," May 10, 1958

In your letter of April 2, 1958, you directed the Secretary of Defense and the Chairman of the National Advisory Committee for Aeronautics to review and report to you which of the "space" programs currently underway or planned by Defense should be placed under the direction of the new civilian space agency proposed in your message to Congress. These instructions specifically stated that "the new Agency will be given responsibility for all programs except those peculiar to or primarily associated with military weapons systems or military operations."

It now appears that the two agencies have reached an agreement contemplating that certain space programs having no clear or immediate military applications would remain the responsibility of the Department of Defense. This agreement would be directly contrary to your instructions and to the concept underlying the legislation the administration has submitted to Congress.

The agreement is primarily the result of the determination of the Defense representatives not to relinquish control of programs in areas which they feel might some day have military significance. The NACA representatives apparently have felt obliged to accept an agreement on the best terms acceptable to Defense.

Specifically, Defense does not wish to turn over to the new agency all projects related to placing "man in space" and certain major component projects such as the proposed million pound thrust engine development. The review by your Scientific Advisory Committee did not see any immediate military applications of these projects.

The effect of the proposed agreement would be to divide responsibility for programs primarily of scientific interest between the two agencies. This would be an undesirable and unnecessary division of responsibility and would be highly impractical. There would not be any clear dividing line, and unnecessary overlap and duplication would be likely. The Bureau of the Budget would have an almost hopeless task in trying to keep the two parts of the program in balance, and problems on specific projects would constantly have to come to you for resolution. The net result of the proposed arrangement would be a less effective program at higher total cost.

On the other hand, it will be relatively simple to work out practical working arrangements under which responsibility and control of the programs in question would clearly be assigned to the new agency as contemplated in your instructions, and the military interest would be recognized by the participation of the Department of Defense in the planning and, where appropriate, the conduct of the programs.

In the circumstances, it is recommended that you direct that the two agencies consult with the Bureau of the Budget and Dr. Killian's office to be sure that any agreement reached is in accordance with the intent of your previous instructions. It is especially important that the announcement of the agreement now being proposed be avoided at this stage of the consideration by the Congress of legislation to establish the new space agency.

While Maxime Faget, a top NACA engineer who would become the "chief designer" of the early U.S. crew-carrying spacecraft, worked with ARPA and the Air Force to plan an initial human spaceflight program, the discussion continued regarding which organization, the Air Force or NASA, should manage that program.

Maxime A. Faget (NACA), Memorandum for Dr. Dryden, June 5, 1958

This memorandum is submitted to review my dealings with ARPA during the past several weeks.

A. Background

    1. I made my first contact with ARPA personnel on May 14, 1958. At that time the NACA was under the impression that it was to work with Dr. Batdorf [an ARPA staff person] to prepare a man-in-space program that would be acceptable to both the NACA and ARPA from a technical standpoint. My first visit to the Pentagon revealed that ARPA had a somewhat different impression of what was to take place. I was told that at the request of Mr. Johnson [Roy Johnson, head of ARPA] a panel had been formed in ARPA to create a man-in-space program and to advise him how this program could best be managed. ARPA had formed this panel approximately a week earlier from members of the ARPA technical staff under the Chairmanship of Dr. Batdorf. I was told that this was only one of many such working groups that was concurrently attacking various jobs in ARPA and that the membership on these various panels greatly overlapped. Accordingly, my position on the man-in-space panel was a special one resulting from an invitation

by Dr. York [Director of Defense Research and Engineering].

2. Inasmuch as this situation was not exactly in keeping with my impression of what it should be, I told the panel that while I would sit as a member of their panel I would also consider myself as a liaison representative of the NACA. In this respect, I reminded them that the direct responsibility for the man-in-space program may quite likely be given to the soon to be created civilian space agency. Thus, I would be concerned that the man-in-space program to be formulated would be one that is acceptable to the NASA and that the management responsibility would be one which could be transferred with the least difficulty. I stated further that if there are any final agreements to be reached between ARPA and NACA they would have to be approved by higher authority, presumably Dr. Dryden and Dr. York and quite possibly by knowledgeable people from the White House. Dr. Batdorf concurred with this and stated that the ARPA staff, most of whom work for IDA [Institute for Defense Analysis], serve only in an advisory capacity . . .

B. Present Situation

1. The work of the panel is apparently drawing to a close. We have put together a proposed man-in-space program that is not far different from the Air Force proposal. The essential elements of this program are:

   a. The system will be based on the use of the Convair Atlas propulsion system. If the expected performance of the Atlas rocket alone is not obtained, then the Atlas 117L system will be used.

   b. The man-in-space flights will be launched from "Pad-20" at AFMTC [Air Force Missile Test Center at Cape Canaveral, Florida].

   c. Retro-rockets will be used to initiate return from orbit.

   d. The non-lifting ballistic type of capsule will be used.

e. The aerodynamic heating during atmospheric entry will be handled by a heat sink or ablation material.

f. Tracking will be carried out primarily with existing or already planned systems. The most important of which will be the G.E. Radio-inertial Guidance System which is highly accurate. The G.E. system will be in existence at AFMTC, San Salvadore, Australia, and Camp Cook.

g. The crew for the orbital flights will be selected from volunteers in the Army, Navy, and Air Force. The crew will be selected in sufficient time to undergo aero-medical training functions.

2. The panel is in unanimous agreement that the man-in-space program should begin immediately. The panel feels that in spite of the unsettled status of both ARPA and NACA that this can be accomplished if a national man-in-space program is adopted. ARPA apparently has $10,000,000 to initiate the program. Future funding and management will of course depend on the outcome of present legislation.

3. The panel is recommending that the Air Force be given the management of the program with executive control to remain in the hands of NACA and ARPA. This could presumably be accomplished by the creation of an executive committee composed of NACA and ARPA people, plus representatives from the contractors, the Air Force, and perhaps Army and Navy.

***

NACA executive director Hugh Dryden chimed in on why the new NASA should be the lead organization for human spaceflight.

Hugh L. Dryden (Director, NACA), Memorandum for James R. Killian Jr., (Special Assistant to the President for Science and Technology), "Manned Satellite Program," July 18, 1958

1. The current objective for a manned satellite program is the determination of man's basic capability in a space

environment as a prelude to the human exploration of space and to possible military applications of manned satellites. Although it is clear that both the National Aeronautics and Space Administration and the Department of Defense should cooperate in the conduct of the program, I feel that the responsibility for and the direction of the program should rest with NASA. Such an assignment would emphasize before the world the policy statement in Sec.102(a) of the National Aeronautics and Space Act of 1958 that "it is the policy of the United States that activities in space should be devoted to peaceful purposes for the benefit of all mankind."

2. The NASA through the older NACA has the technical background, competence, and continuing within-government technical back-up to assume this responsibility with the co-operation and participation of the Department of Defense. For a number of years, the NACA has had groups doing research on such items as stabilization of ultra-high-speed vehicles, provision of suitable controls, high-temperature structural design, and all the problems of reentry. More recently, the NACA research groups have been working on these problems with direct application to manned satellites. The human-factors problems of this program are not far different from those for the X-15 which the NACA has been studying in cooperation with the Navy and Air Force. Thus, the NACA has enlisted the cooperation of the military services and marshaled the required technical competence. Included in this competence are large, actively-working, staffs in NACA laboratories providing additional technical back-up for the manned-satellite program.

3. The assignment of the direction of the manned satellite program to NASA would be consistent with the President's message to Congress and with the pertinent extracts from the National Aeronautics and Space Act of 1958 given in the appendix to this memorandum.

In August 1958 President Eisenhower ended the controversy by officially assigning the human spaceflight mission to NASA. When NASA opened, Administrator T. Keith Glennan quickly approved a plan for a piloted satellite project and on October 8 created the Space Task Group (STG), based at NASA's Langley Research Center in Hampton, Virginia, to manage the project. Heading STG would be Robert Gilruth, a veteran NACA engineer and manager. Thirty-five people from Langley and ten from NASA's Lewis Research Center in Cleveland were transferred to the STG; they would form the nucleus of some one thousand people eventually working on the project.

The new project needed a name. The space agency chose "Project Mercury," after the messenger of the gods in ancient Roman mythology. The symbolic associations of this name appealed to Abe Silverstein, NASA's director of space flight development. On December 17, 1958, the fifty-fifth anniversary of the first flight of the Wright brothers at Kitty Hawk, Glennan was scheduled to announce the name. At the last minute, the official responsible for human spaceflight at NASA Headquarters, George Low, on behalf of the Space Task Group's Robert Gilruth, attempted to change the name. He was not successful. The first U.S. human spaceflight effort would go down in history as Project Mercury.

Abe Silverstein (Director of Space Flight Development, NASA), Memorandum for Administrator (NASA), "Code Name 'Project Mercury' for Manned Satellite Project," November 26, 1958

1. Considerable confusion exists in the press and in public discussions regarding the Manned Satellite Project because of the similarity of this program with other Man-in-Space proposals.

2. At the last meeting of the Manned Satellite Panel it was suggested that the Manned Satellite Project be referred to as Project Mercury.

3. It is recommended that the code name Project Mercury be adopted.

George M. Low (NASA), Memorandum for Dr. Silverstein (NASA), "Change of Manned Satellite Project Name from 'Project Mercury' to 'Project Astronaut,'" December 12, 1958

1. Bob Gilruth feels that "Project Astronaut" is a far more suitable name for the Manned Satellite Project than "Project Mercury."
2. If you agree, this should be brought to Dr. Glennan's attention immediately. Present plans call for Dr. Glennan to refer to "Project Mercury" in his policy speech on December 17.

In fall 1958, Space Task Group engineers, led by Max Faget, set out designing the Mercury spacecraft and selecting an industrial contractor to construct it. NASA received eleven proposals in December 1958 from firms wanting the spacecraft development and production contract, and in January 1959 selected McDonnell Aircraft, based in St. Louis, as the Mercury spacecraft prime contractor.

There remained the not-insignificant task of identifying the type of person who would ride the Mercury spacecraft into space, and then selecting the individuals who met the requirements of that position, initially designated "Research Astronaut." In November 1958 aeromedical consultants working for the Space Task Group worked out preliminary procedures for the selection of astronauts to pilot the Mercury spacecraft. Their proposal anticipated meetings with the relevant industrial firms and the military services to solicit nominees; the target was the nomination of 150 men. (Women were specifically not invited to apply.) This would be narrowed down to thirty-six people to undergo extensive physical and psychological testing. Ultimately, twelve men would be selected to undergo training and qualification, of which only six were expected to fly.

This plan led Charles Donlan, technical assistant to the director of Langley; Warren J. North, a former NACA test pilot and head of Langley's Office of Manned Satellites; and Allen O. Gamble, a psychologist detailed from the National Science Foundation, to draft job specifications for applicants for the astronaut program. NASA at first intended to hold an open competition for the position, and on December 22, 1958, issued an invitation to apply. Even though Glennan had already an-

nounced that the program would be called "Project Mercury," the application uses the name preferred by the Space Task Group, "Project Astronaut."

NASA, "Invitation to Apply for Position of Research Astronaut-Candidate, NASA Project A, Announcement No. 1," December 22, 1958

Invitation to apply for Position of <u>RESEARCH ASTRONAUT-CANDIDATE</u> with minimum starting salary range of $8,330 to $12,770 (GS-12 to GS-15) depending upon qualifications at the NASA Langley Research Center Langley Field, Virginia

I. DESCRIPTION OF PROJECT ASTRONAUT

The Manned Satellite Project is being managed and directed by NASA. The objectives of the project are to achieve, at the earliest practicable date, orbital flight and successful recovery of a manned satellite; and to investigate the capabilities of man in a space environment. To accomplish these objectives, a re-entry vehicle of the ballistic type has been selected. This vehicle not only represents the simplest and most reliable configuration, but has the additional advantage of being sufficiently light, so that it can be fitted on an essentially unmodified ICBM booster. The satellite will have the capability of remaining in orbit for 24 hours, although early flights are planned for only one or two orbits around the earth.

Although the entire satellite operation will be possible, in the early phases, without the presence of man, the astronaut will play an important role during the flight. He will contribute to the reliability of the system by monitoring the cabin environment, and by making necessary adjustments. He will have continuous displays of his position and attitude and other instrument readings, and will have the capability of operating the reaction controls, and of initiating the descent from orbit. He will contribute to the operation of the communications system. In addition, the astronaut will make research observations that cannot be made by instruments; these include physiological, astronomical, and meteorological observations.

Orbital flight will be accomplished after a logical buildup of capabilities. For example, full-scale capsules will be flown on short and medium range ballistic flights, before orbital flights will be attempted. Maximum effort will be placed on the design and development of a reliable safety system. The manned phases of the flight will also undergo a gradual increase in scope, just as is common practice in the development of a new research aircraft.

II. DUTIES OF RESEARCH ASTRONAUT-CANDIDATES

Research Astronaut-Candidates will follow a carefully planned program of pre-flight training and physical conditioning. They will also participate directly in the research and development phase of Project Astronaut, to help insure scientifically successful flights and the safe return of space vehicles and their occupants. The duties of Research Astronaut-Candidates fall into three major categories:

a. Through training sessions and prescribed reading of technical reports, they will acquire specialized knowledge of the equipment, operations, and scientific tests involved in manned space flight. They will gain knowledge of the concepts and equipment developed by others and, as their knowledge and experience develops, they will contribute their thinking toward insuring maximum success of the planned flights.

b. They will make tests and act as observers—under-test in experimental investigations designed (1) to develop proficiency and confidence under peculiar conditions such as weightlessness and high accelerations; (2) to enable more accurate evaluation of their physical, mental, and emotional fitness to continue the program; and (3) to help elicit the knowledge necessary to evaluate and enable the final development of communication, display, vehicle-control, environmental-control, and other systems involved in space flight.

c. They will perform special assignments in one or more of their areas of scientific or technical competence, as an adjunct to the regular programs of the research team, the research center, or NASA.

d. These assignments may include doing research, directing or evaluating test or other programs, or doing other work which makes use of their special competencies.

Appointees who enter this research and training program will be expected to agree to remain with NASA for 3 years, including up to one year as Research Astronaut-Candidates. During the initial months final selection will be made of about half of the group to become Research Astronauts. Candidates who are not at that point designated Research Astronauts will have the option of continuing with NASA in other important capacities which require their special competence and training, without loss of salary and with other opportunities for advancement, and may remain eligible for future flights.

III. QUALIFICATION REQUIREMENTS

A. Citizenship, Sex, Age

Applicants must be citizens of the United States. They must be males who have reached their 25th birthday but not their 40th birthday on the date of filing application.

Applicants must be in excellent condition and must be less than 5 feet 11 inches in height.

B. Basic Education

Applicants must have successfully completed a standard 4-year or longer professional curriculum in an accredited college or university leading to a bachelor's degree, with major study in one of the physical, mathematical, biological, medical, or psychological sciences or in an appropriate branch of engineering or hold a higher degree in one of these fields . . .

C. Professional Experience or Graduate Study

In addition to a degree in science or engineering or medicine, applicants must have had one of the following patterns of professional work or graduate study or any equivalent combination:

1. Three years of work in any of the physical, mathematical, biological, or psychological sciences.
2. Three years of technical or engineering work in a research and development program or organization.
3. Three years of operation of aircraft or balloons or submarines, as commander, pilot, navigator, communications officer, engineer, or comparable technical position.
4. Completion of all requirements for the Ph.D. degree in any appropriate field of science or engineering, plus 6 months of professional work.
5. In the case of medical doctors, 6 months of clinical or research work beyond the license and internship or residency.

Preference will be given to applicants in proportion to the relatedness of their experience or graduate study to the various research and operational problems of astronautics.

NASA desires to select and train a team of Astronaut-Candidates representing a variety of fields including physical and life sciences and technology.

D. Hazardous, Rigorous, and Stressful Experience

Applicants must have had a substantial and significant amount of experience which has clearly demonstrated three required characteristics: (a) willingness to accept hazards comparable to those encountered in modern research airplane flights; (b) capacity to tolerate rigorous and severe environmental conditions; and (c) ability to react adequately under conditions of stress or emergency.

These three characteristics may have been demonstrated in connection with certain professional occupations such as test pilot, crew member of experimental submarine, or arctic or antarctic explorer. Or they may have been demonstrated during wartime combat or military training. Parachute jumping or mountain climbing or deep sea diving (including with SCUBA), whether as occupation or sport, may have provided opportunities for demonstrating these characteristics, depending upon heights or depths attained, frequency and duration, temperature and other environmental conditions, and emergency episodes encountered. Or they may have been demonstrated by experience as an observer-under-test for extremes of environmental conditions such as acceleration, high or low atmospheric pressure, variations in carbon dioxide and oxygen concentration, high or low ambient temperatures, etc. Many other examples could be given. It is possible that the different characteristics may have been demonstrated by separate types of experience.

Pertinent experience which occurred prior to 1950 will not be considered. At least some of the pertinent experience must have occurred within one year preceding date of application.

Applicants must submit factual information describing the work, sport, or episodes which demonstrate possession of these three required characteristics.

***

V. SELECTION PROGRAM

On the basis of evaluations of the above-described applications and supporting material, a group of men will be invited to report to the NASA Space Task Force at Langley Field, Virginia, on February 15, 1959. For about three weeks these men will be given a variety of physical and mental tests on a competitive basis to evaluate their fitness for training for the planned space flights. This will involve trips to Washington, D.C., and other locations and will include tests with such equipment as decompression chambers and centrifuges and also aircraft flights. At the end of this competitive testing program all the candidates will return to their homes and jobs.

During the ensuing period of 2 to 3 weeks, laboratory and other test results will be evaluated and a small group of men will be finally selected to become Research Astronaut-Candidates. These men will be notified to report for duty at NASA, Langley Field, on or about April 1, 1959. Travel and moving expenses for them (and their families, if married) will be provided.

VI. APPOINTMENTS AND PAY

These appointments are to civilian positions in the National Aeronautics and Space Administration. They are excepted appointments due to the unusual nature of the duties and the selection process, but carry the benefits and protections of the

U.S. Civil Service System including a high level of insurance and retirement.

Original appointments of Research Astronaut-Candidates will be to pay levels commensurate with their backgrounds of education and experience, within the pay range of $8,330 to $12,770 per year (GS-12 to GS-15).

As these men become proficient in the field, they will become eligible for Research Astronaut positions with salaries commensurate with those of the most highly skilled NASA Research Pilots and Aeronautical and Space Scientists.

The plan to solicit astronaut applications from both industry and the military was soon abandoned; instead, President Eisenhower decided that only military test pilots should be allowed to apply. This both simplified the recruitment process and made sure that the first U.S. astronauts would be male, since there were no female military test pilots. After this presidential directive, NASA screened military personnel records and found 110 men who met the minimum requirements. NASA made initial contacts with some of those men, and thirty-two were selected for a rigorous set of physical and mental examinations. The original intent had been to select up to twelve astronaut candidates, to allow for those who might drop out as Project Mercury progressed, but the level of motivation among the thirty-two semifinalists was so high that NASA decided it needed to select only six people. No applicants from the Army made the final cut; NASA made a second search to see if a qualified Army officer was available, but without success. Unable to decide which six candidates were best qualified, NASA ultimately chose seven men: one Marine, three from the Navy, and three from the Air Force. They were Scott Carpenter, Gordon Cooper, John Glenn, Virgil "Gus" Grissom, Walter Schirra, Alan Shepard, and Donald "Deke" Slayton. As America's first astronauts, they would soon come to be known as the "Mercury 7."

The newly selected astronauts were introduced to the country at an exuberant April 9, 1959, NASA press conference. They became instantaneous celebrities, with insistent clamor for every detail of their lives. But it was still unknown whether these astronauts could actually survive a trip to space. In another step to understand what would happen when humans were launched into space and to test the life-support systems of the Project Mercury spacecraft, NASA decided to carry out a series of test flights launching primates—monkeys and chimpanzees—on various trajectories. The Soviet Union was at the same time launching animals into space, employing dogs as test subjects. The use of primates was thought to provide a better simulation of the impact of spaceflight on a human, but NASA also realized that the human-like attributes of primates meant that their use had to be carefully managed to avoid criticism of cruelty to animals.

There were four such launches. Two used the "Little Joe" booster, a low-cost solid-fueled rocket. The other two launches used the same boosters that would be used for crew-carrying flights. The first launch was the Little Joe 2 flight of December 4, 1959, with Sam, an American-born rhesus monkey, aboard. Sam was recovered several hours later, with no ill effects from his journey. He was later returned to his home at the School of Aviation Medicine at Brooks Air Force Base, San Antonio, Texas. Miss Sam, another rhesus monkey and Sam's mate, was launched on January 21, 1960, on a second Little Joe mission. She was also recovered and returned to the School of Aviation Medicine.

Ham, a chimpanzee whose name was an acronym for Holloman AeroMed, became the first primate in space on the suborbital Mercury Redstone 2 mission. The flight was launched on January 31, 1961; it carried Ham on a 16-minute, 39-second flight that, due to the Redstone booster not operating properly, traveled downrange 124 miles more than had been planned and subjected Ham to some seventeen times the force of gravity, well above the anticipated amount. Even so, Ham survived the flight in good health.

The chimpanzee Enos became the first primate to orbit the Earth, on the Mercury Atlas 5 mission launched on November 29, 1961. This two-orbit flight proved the capability of the Mercury spacecraft for orbital flight and cleared the way for John Glenn's February 1962 first U.S. crewed orbital flight. Because of the interest in and the sensitivity about these primate flights, NASA took considerable pains to explain how the animals were to be treated and what role they played in the program.

NASA, "Information Guide for Animal Launchings in Project Mercury," July 23, 1959

1. BACKGROUND

Animals will be used in the Project Mercury developmental program to gain information on the biological response to space flight. Problems facing manned orbital flight essentially are engineering in nature, and the animal program will be relatively simple in scope. Knowledge from animal flights will contribute information to the program in the areas of life support systems; instruments to measure physiological reactions in the space environment; prove out design concepts when they are near known limits in such areas as high-g loads; test equipment and instrumentation under dynamic load conditions, and to develop countdown procedures and train personnel in these procedures prior to manned flight.

NASA has selected three animals for developmental work in the Mercury program: rhesus monkey (Macaca Mulatta), chimpanzee and mouse. Primates were chosen because they have the same organ placement and suspension as man. Both rhesus and chimp have relatively long medical research backgrounds, and the type of rhesus born and bred in American vivarium has a 20-year research background as a breed. The chimp is larger and more similar to man in body systems, and will be used for advanced developmental flights with the McDonnell Aircraft Corporation capsule.

***

NASA Washington will be the source of all post-launch scientific information. A Technical press briefing will be conducted about 24 hours after the launch to summarize all information known at that time. Representatives on the panel will be from NASA Headquarters; Space Task Group; STG Biomedical Group; Launching Team, and Recovery Team.

Pre-launch information activities require training and housing still and motion pictures of the animal subjects. The responsible agency (i.e., USAF) will be asked to provide footage and a selection of photographs in these areas.

NASA will take and provide photographs (still and motion

picture) of the animals in biopacks and the biopack insertion into the capsule.

At no time will the animal subjects be available to the press either for photography or viewing. NASA will follow this policy for these reasons:

1   Test results are influenced by excitement, particularly since the animal subjects have led sheltered lives. A minimum of crowd activity is justified from both scientific and clinical standpoints.

2   Elimination of all but necessary scientific persons curtails added chances of the primates contacting diseases.

3   Complex handling procedures for the animals will not be required.

4   The undesirable effects of the "Roman Holiday" atmosphere are eliminated.

***

Each of the Mercury 7 was a "type A" personality, oriented toward action and achievement. Recognizing that their position gave them potential political leverage, the seven soon after their selection suggested that they engage in "astronaut diplomacy" with their Soviet counterparts, but that initiative was not welcomed in Washington. The first meetings between U.S. astronauts and Soviet cosmonauts did not take place until 1962, after both countries had made orbital flights. In this memo, the Mercury astronauts carefully weigh the pros and cons of an early astronaut-cosmonaut meeting.

Mercury Astronauts, Memorandum for NASA Project Director, "Exchange of Visits with Russian Astronauts," October 21, 1959

1.   The Russians have recently announced their man-in-space program and have given some publicity to the pilots selected. In the eyes of the rest of the world, it appears that Project Mercury is placed in a competitive position, whether we like it or not. This, of course, sets up for another barrage of unfavorable propaganda when, and if, the Russians achieve space flight before we do.

2. Certain action at this time might place us in a better position to gain information about their program and also take the propaganda initiative away from the Russians with regard to manned space flight. Suggested action is to propose mutual visits between the Astronauts of the two countries with the purpose of sharing information on training and mutual problem areas.

3. Propaganda-wise, we apparently stand to gain a great deal and could lose little or nothing.

   a. The U.S. would have taken the initiative in sponsoring international cooperation in the manned space field.

   b. Such a proposal would support, to the world, our statements of the peaceful intent of Project Mercury as a scientific exploration with no ulterior motives.

   c. It is in keeping with the current political atmosphere engendered by the Khrushchev visit and the proposed presidential visit to Russia.

4. There appears to be little we could lose, in that practically all of the details of Project Mercury are already public domain and have been covered repeatedly in the press. The Russian program, on the other hand, has been secret, so anything we could learn would be new information.

5. Refusal of the Russians to cooperate in such a proposal would certainly reflect unfavorably in the eyes of other countries. These are countries already concerned about where the American-Russian space race is leading.

6. Timing of such a proposal is very important. If such a proposal is made, it should be done very soon, before either Russia or the U.S. has accomplished a man-in-space mission.

7. If we wait until we make the first orbital flight, and then propose an exchange, it would appear that we are "rubbing it in" a little and are willing to throw a little information to our poor cousins who could not do it themselves. This would probably do us more harm than good in the attitude with the rest of the world.

8. If, on the other hand, we wait until the Russians have made the first orbital flight before we propose such an exchange, it would appear that we are trying to get information on how they did it because we have not been able to do the

same thing. This would also do us harm in the eyes of other countries.

9. To summarize, we stand to gain information in an exchange of visits, while giving little information that is not already known. Propaganda value of such a proposal and visit should be very favorable for us, if the proposal is made from the U.S. and before either country has made an orbital flight.

10. One way to assess the value of such a proposal is to think of our reaction and the reaction of other countries if the Russians make such a proposal first. It appears that we stand to gain by making the proposal first.

11. It is realized that there are many considerations involved in such a proposal. NASA, State Department, Intelligence, and many other government sources concerned must have vital inputs that will determine whether the proposal is not only feasible, but advisable.

12. The proposal is herewith submitted for consideration.

Away from the press, the Mercury 7 began to train for their unprecedented role as space flyers. Project Mercury was basically a straightforward experiment to learn what would happen to a human being when subjected to the high-gravity loads of launch, a rapid transition to weightlessness, and then the return to gravity during a stressful return to Earth from space. It was not known whether a person could survive that experience, and, if he did, whether he could perform various functions during a spaceflight. Given the unknowns related to astronaut reaction to spaceflight, the Mercury spacecraft was designed to function automatically, with astronaut control only as a backup. Even so, NASA during 1959–60 developed a training regime for the astronauts that attempted to simulate as closely as possible the stresses of spaceflight. The seven were given training on all spacecraft systems and indeed on any aspect of the spaceflight experience that NASA thought might be relevant. The training regime was intense and comprehensive.

This letter from astronaut and Marine Lieutenant Colonel John Glenn to then Lieutenant Commander James B. Stockdale, United States Navy, offers a personal perspective on the early

Mercury program. Glenn describes various aspects of astronaut training, including being subjected to forces sixteen times Earth gravity (16 gs). Stockdale, a classmate of Glenn's at the Navy's Test Pilot School at Patuxent River, Maryland, would later gain fame as one of the earliest heroes of Vietnam. He was shot down in 1965 and was held as a prisoner of war for seven years. In 1992, Stockdale was a candidate for vice president of the United States on a ticket headed by businessman Ross Perot.

Letter from John Glenn to Lieutenant Commander Jim Stockdale, USN, December 17, 1959

Dear Jim:

***

This past 8 or 9 months has really been a hectic program to say the least and by far the most interesting thing in which I have ever taken part, outside of combat. It is certainly a fascinating field, Jim, and growing so fast that it is hard to keep up with the major developments, much less everything in the field.

Following our selection in April, we were assigned to the Space Task Group, portion of NASA at Langley Field, and that is where we are based when not traveling. The way it has worked out, we have spent so much time on the road that Langley has amounted to a spot to come back to get clean skivvies and shirts and that's about all. We have had additional sessions at Wright Field in which we did heat chamber, pressure chamber, and centrifuge work and spent a couple of weeks this fall doing additional centrifuge work up at NADC [Naval Air Defense Center], Johnsville, Pennsylvania. This was some program since we were running it in a lay-down position similar to that which we will use in the capsule later on and we got up to as high as 16 g's. That's a batch in any attitude, lay-down or not.

With the angles we were using, we found that even lying down at 16 g's, it took just about every bit of strength and technique you could muster to retain consciousness. We found there was quite a bit more technique involved in taking this kind of g than we had thought. Our tolerances from beginning to end of runs during the period we worked up there went up considerably as we developed our own technique for taking this high g. A few runs a day like that can really get to you. Some other stuff we did up there involved what we call tumble runs or going from a +g in 2 seconds to a −g and the most we did on this was in going from a +9g to a −9g. Obviously, a delta of 18 . . . When we first talked about doing this, I didn't think it would be possible at all, but in doing a careful build-up, we happily discovered that this was not so horrible. At +9g to −9g, we were bouncing around a bit but it was quite tolerable.

I guess one of the most interesting aspects of the program has been in some of the people we have been fortunate enough to meet and be briefed by. One of the best in this series was the time we spent at Huntsville, Alabama, with Dr. Wernher von Braun and crew. We were fortunate enough to spend an evening with him in his home until about 2:30 in the morning going through a scrap book, etc., from Peenemunde days in Germany and, in general, shooting the bull about his thoughts on the past, present, and future of space activities. This was a real experience for a bunch of country boys fresh caught on the program and a very heady experience as you can imagine.

We have had a good run-down at Cape Canaveral and got to see one of their shots. I guess that is one of the most dramatic things I have ever seen. The whole procedure they go through for a night launch at the Cape is just naturally a dramatic picture far better than anything Hollywood could stage. When the Big Bird finally leaves the pad, it doesn't have to be hammed up to be impressive.

Much of our work, of course, has involved engineering work on the capsule and systems. My particular specialty area has been the cockpit layout and instrumentation presentation for the Astronauts. This has been extremely interesting because we are working on an area way out in left field where our ideas are as good as any one else's. So, you try to take the best of your past experiences and launch from there with any new ideas you can contrive. This is the kind of development work, as you well know, that is by far the most enjoyable.

We just finished an interesting activity out at Edwards Air Force Base doing some weightless flying in the F-100. This was in the two-place F-100 so that we could ride in the rear seat and try various things such as eating and drinking and mechanical procedures while going through the approximately 60-second ballistic parabola that you make with the TF-100. That started at about 40,000 feet, 30-degree dive to 25,000, picking up about 1.3 to 1.4

Mach number, pull out and get headed up hill again at 25,000 and about a 50-to-60-degree climb angle, at which point they get a zero-g parabola over the top to about 60 degrees down hill.

You can accomplish quite a bit in a full minute in those conditions and contrary to this being a problem, I think I have finally found the element in which I belong. We had done a little previous work floating around in the cabin of the C-131 they use at Wright Field. That is even more fun yet, because you are not strapped down and can float around in the capsule doing flips, walk on the ceiling, or just come floating full length of the cabin while going through the approximately 15-seconds of weightlessness that they can maintain on their shorter parabola. That was a real ball and we get some more sessions with this machine some time after the first of the year.

Before this next year is out, we should get the manned Redstone ballistic shots started which will put us to orbital altitude of 105 nautical miles, but not up to the orbital speeds so that we arc back down off the Cape about 200 miles from the pad. We figure now that the first actual manned orbital shots should follow in mid to late 1961.

***

In preparing for the initial U.S. human spaceflights, the seven Mercury astronauts engaged in weightless simulations during parabolic flight, centrifuge tests, altitude chamber research, physical fitness training, survival school, pilot proficiency preparation, and a host of other activities. This report, prepared by one of the people who designed the astronaut training program, explains how these various activities fit together. The complexity and comprehensiveness of the training program suggest the intense character of the Project Mercury experience.

Robert B. Voas (NASA Space Task Group), "Project Mercury Astronaut Training Program," May 26, 1960

***

TRAINING PROGRAM

The Astronaut training program can be divided into six major topic areas. The primary requirement, of course, is to train the

Astronaut to operate the vehicle. In addition, it is desirable that he have a good background knowledge of such scientific areas related to space flight as propulsion, trajectories, astronomy, and astrophysics. He must be exposed to and familiarized with the conditions of space flight such as acceleration, weightlessness, heat, vibration, noise, and disorientation. He must prepare himself physically for those stresses which he will encounter in space flight. Training is also required for his duties at ground stations before and after his own flight and during the flight of other members of the Astronaut team. An aspect of the training which might be overlooked is the maintenance of the flying skill which was an important factor in his original selection for the Mercury program.

Training in vehicle operation. Seven training procedures or facilities were used in developing skills in the operation of the Mercury capsule. These included lectures on the Mercury systems and operations; field trips to organizations engaged in the Mercury Project; training manuals; specialty study programs by the individual Astronaut; mockup inspections; and training devices. To provide the Astronaut with a basic understanding of the Mercury system, its components, and its functions, a lecture program was set up. A short trip was made to McDonnell at which time a series of lectures on the capsule systems was presented. These systems lectures were then augmented by lectures on operations areas by Space Task Group scientists. This initial series of lectures provided a basis for later self-study, in which use was made of written descriptive material as it became available. Individual lectures have been repeated as the developments within Project Mercury have required a series of lectures on capsule systems by both Space Task Group and McDonnell personnel. These lectures have been scheduled to coincide with the delivery and initial operation of the fixed-base Mercury trainer. In these lectures, the same areas are reviewed in an attempt to bring the Astronauts up-to-date on each of the systems as they begin their primary procedures training program.

In addition to this lecture program, indoctrination trips have been made to the major facilities concerned with the Project Mercury operations. Two days were spent at each of the following facilities: McDonnell, Cape Canaveral, Marshall Space Flight Center, Edwards Flight Test Center, and Space Technology

Laboratories and Air Force Ballistic Missile Division. One day was spent at Rocketdyne Division, North American Aviation, and five days were spent at Convair/Astronautics. At each site there was a tour of the general facilities together with a viewing of Mercury capsule or booster hardware and lectures by top-level personnel covering their aspect of the Mercury operation. The Astronauts also had an opportunity to hear of related research vehicles such as the X-15 and Discoverer and received a brief discussion of the technical problems arising in these programs and their significance to Project Mercury.

***

Another method employed to aid in the dissemination of information to the Astronauts was to assign each a specialty area. These assignments were as follows: M. Scott Carpenter, navigation and navigational aids; Leroy G. Cooper, Redstone booster; John A. Glenn, crew space layout; Virgil I. Grissom, automatic and manual attitude control system; Walter M. Schirra, life support system; Alan B. Shepard, range, tracking, and recovery operations; and Donald K. Slayton, Atlas booster. In pursuing these specialty areas, each man attends meetings and study groups at which current information on capsule systems is presented. Regular periods are set aside for all the men to meet and report to the group. Another important source of information about the vehicle, particularly in the absence of any elaborate fixed-base trainers, has been the manufacturer's mockup. Each of the men has had an opportunity to familiarize himself with the mockup during visits to McDonnell.

Following the initial familiarization with the Mercury system, the primary training in vehicle operation is being achieved through special training devices developed for the Mercury program. Early training in attitude control was accomplished on the Langley Electronics Associates Computer . . . which was combined with a simulated Mercury attitude display and hand controller. This device was available during the summer of 1959. Later, another analog computer was cannibalized from an F-100F simulator and combined with actual Mercury hardware to provide more realistic displays and controls. This MB-3 trainer . . . also included provision for the Mercury couch and the pressure suit.

In addition to these two fixed-base simulators, three dynamic simulators were used to develop skill in Mercury attitude control. The first, of these, the ALFA (Air Lubricated Free Attitude) Simulator . . . permits the practice of orbit and retrofire attitude control problems by using external reference through simulated periscope and window displays. A simulated ground track is projected on a large screen which is viewed through a reducing lens to provide the periscope display. This simulator also permits training in the use of earth reference for navigation. The Johnsville Centrifuge . . . was used as a dynamic trainer for the reentry rate damping task because it adds the acceleration cues to the instruments available in the fixed-base trainers. It also provides some opportunity to practice sequence monitoring and emergency procedures during launch and reentry. Another dynamic simulation device used to provide training in recovery from tumbling was the three-gimbaled MASTIF (Multi-Axis Spin Test Inertia Facility) device at the NASA Lewis Laboratory . . . In this device, tumbling rates up to 30 rpm in all three axes were simulated and the Astronaut was given experience with damping these rates and bringing the vehicle to a stationary position by using the Mercury rate indicators and the Mercury-type hand controller.

Two more elaborate trainers became available in the summer of 1960. These trainers provide practice in sequence monitoring and systems management. The McDonnell Procedures Trainer . . . is similar to the fixed-base trainers which have become standard in aviation operations. The computer used on the MB-3 has been integrated with this device to provide simulation of the attitude control problem. External reference through the periscope is simulated by using a cathode ray tube with a circle to represent the earth. Provision has been made for pressurizing the suit and for some simulation of heat and noise effects. The environmental control simulator . . . consists of the actual flight environmental control hardware in the capsule mockup. The whole unit can be placed in a decompression chamber in order to simulate the flight pressure levels. This device provides realistic simulation of the environmental-control system functions and failures. Effective use of these two simulators is predicated upon adequate knowledge of the types of vehicle systems malfunctions which can occur. A failure-mode analysis carried out by the manufacturer has provided a basis for determining the types of malfunction which are possible and the requirements for simulating them . . .

A record system on which possible malfunctions are listed on cards, together with methods of their simulation, has been set up. On the back of these cards there is space for noting when and under what conditions this failure has been simulated and what action the Astronaut took to correct it. In this way, it is hoped that the experience in the detection and correction of systems malfunctions can be documented.

Training in space sciences. In addition to being able to operate the Mercury vehicle, the Astronaut will be required to have a good general knowledge of astronomy, astrophysics, meteorology, geophysics, rocket engines, trajectories, and so forth. This basic scientific knowledge will enable him to act as a more acute observer of the new phenomena with which he will come in contact during the flight. It will also provide a basis for better understanding of the detailed information which he must acquire on the Mercury vehicle itself. In order to provide this broad background in sciences related to astronautics, the Training Section of the Langley Research Center set up a lecture program which included the following topics: Elementary Mechanics and Aerodynamics (10 hours); Principles of Guidance and Control (4 hours); Navigation in Space (6 hours); Elements of Communication (2 hours); Space Physics (12 hours). In addition, Dr. W. K. Douglas, Flight Surgeon on the Space Task Group staff, gave 8 hours of lectures on physiology.

Following this initial lecture program, training in specific observational techniques is planned. The first activity of this program was training in the recognition of the primary constellations of the zodiac at the Morehead Planetarium in Chapel Hill, North Carolina. A Link trainer body was modified with a window and headrest to simulate the capsule external viewing conditions. Using this device, the Astronauts were able to practice the recognition of constellations which the Planetarium was programmed to simulate orbital flight. Future plans call for further training in star recognition together with methods of observing solar and meteorological events, earth and lunar terrain, and psychological and physiological reactions. These activities will be in support of a primary objective of the Project Mercury program which is to determine man's capability in a space environment. The training program contributes to this objective in three ways:

1. First, by establishing base lines, both for the Astronaut's performance and his physiological reactions. These base lines can then be compared with psychological and physiological factors in the space environment.

2. Second, through the program in basic sciences described above, the Astronaut is given sufficient background with which to appreciate the importance of the observations which he can make in the space environment.

3. Specific training in observational techniques and the use of scientific equipment arms him with the skills with which to collect data of value to science.

Thus, the training program attempts to lay the ground work for the scientific activities of the Astronauts, as well as to provide the specific skills which are required to fly the Mercury vehicle.

Familiarization with conditions of space flight. An essential requirement of the training program is to familiarize the Astronaut with the novel conditions which man will encounter in space flight. An important part of the Astronaut training program has been to provide the trainees with an opportunity to experience eight types of conditions associated with Mercury flights: high acceleration, weightlessness, reduced atmospheric pressures, heat, disorientation, tumbling, high concentration of CO_2, and noise and vibration.

The Astronauts experienced acceleration patterns similar to those associated with the launch and reentry of the Mercury [spacecraft] first at the Wright Air Development Division (WADD) in Dayton, Ohio, and later at the Aviation Medical Acceleration Laboratory at Johnsville, Pennsylvania. During this training, they were able to develop straining techniques which reduced the problem of blackout and chest pain. It was generally the opinion of the Astronauts that the centrifuge activity was one of the most valuable parts of the training program.

The Astronauts were given an opportunity to experience weightless flying both in a free-floating condition in C-131 and C-135 aircraft and strapped down to the rear cockpit of a F-100 F fighter. While the latter is more similar to the Mercury operation, the Astronauts, being experienced pilots, felt that there was little or no difference between this experience and their normal flying activities. The free-floating state, however, they felt was a novel and enjoyable experience. Since the longer period of weightlessness available in the F-100 F aircraft is valuable for collecting medical data, while the C-131 aircraft appears to give the most interesting experiential training, both types of operations appear to be desirable in a training program. The fact that the pilots experienced no unusual sensations during weightlessness when fully restrained was an encouraging finding for the Mercury operation and supports the desirability of selecting flying personnel for this type of operation.

***

Physical fitness program. To insure that the Astronaut's performance does not deteriorate significantly under the various types of stresses discussed in the previous section, it is important that he be in excellent physical condition. Since most of the trainees entered the Project Mercury program in good physical health, a group physical fitness program, with one exception, has not been instituted. SCUBA training was undertaken because it appeared to have a number of potential benefits for the Project Mercury, in addition to providing physical conditioning. It provides training in breathing control and analysis of breathing habits, and in swimming skill (desirable in view of the water landing planned in the Mercury program). Finally, there is, in the buoyancy of water, a partial simulation of weightlessness, particularly if vision is reduced. Aside from this one organized activity, each individual has been undertaking a voluntary fitness program tailored to his own needs. This program has included, for most of the Astronauts, three basic items. First of all, as of December 1959, they have reduced or completely stopped smoking. This was an individual, voluntary decision and was not a result of pressure by medical personnel, but a result of their own assessment of the effect of smoking on their tolerance to the stresses to be encountered in the flight, particularly acceleration. Some of the members of the team who have a tendency to be overweight have initiated weight-control programs through proper diet. Nearly all members make it a habit to get some form of daily exercise.

Training in ground activities. Frequently overlooked are the extent and the importance of the ground activities of the Astronauts. Their knowledge of the vehicle and its operation makes them specially qualified for certain ground operations. The training in ground procedures has fallen into the three main areas; countdown procedures, ground flight monitoring procedures, and recovery and survival. The Astronauts are participating in the development of the countdown procedures and will be training themselves in their own part of the countdown through observation of countdown procedures for the initial unmanned shots, and finally, by participating in the preparation procedures for the actual manned flights.

**

Maintenance of flight skills. One of the continuing problems in training for space flight is the limited opportunity for actual flight practice and proficiency training. The total flight time in the Mercury capsule will be no more than 4 to 5 hours over a period of 3 years for each Astronaut. The question arises as to whether all the skills required in operating the Mercury vehicle can be maintained purely through ground simulation. One problem with ground simulation relates to its primary benefit. Flying a ground simulator never results in injury to the occupant or damage to the equipment. The penalty for failure is merely the requirement to repeat the exercise. In actual flight operations, failures are penalized far more severely. A major portion of the Astronaut's tasks involves high-level decision making. It seems questionable whether skill in making such decisions can be maintained under radically altered motivational conditions. Under the assumption that vigilant decision making is best maintained by experience in flight operations, the Mercury Astronauts have been provided with the opportunity to fly high-performance aircraft. The program in this area is a result of their own interest and initiative and is made possible by the loan and maintenance of two F-102 aircraft by the Air Force.

IMPLICATIONS FOR FUTURE PROGRAMS

In conclusion, the problems with implications for future space flight projects which have been encountered in development of the Mercury program can be reviewed.

In developing skills in operation of the vehicle, the difficulty of providing up-to-date information on the systems when the training must progress concurrently with the development program has been discussed. Concurrent training and development should tend to be a feature of future space flight programs, since many of these will be experimental in nature, rather than operational.

All spacecraft have in common the problem of systems which must be kept functional for long periods without recourse to ground support. Even in the event of emergency termination of the mission with immediate return to earth, prolonged delay may occur before safe conditions within the atmosphere have been achieved. Thus, emphasis on "systems management" will increase in future space operations. Recognition of malfunctions has always been a part of the pilot's task; usually, however, little in-flight maintenance is attempted. Since aborts are dangerous and, in any event, involve greater delay before return, the Astronaut must do more detailed diagnoses of malfunctions and more in-flight maintenance. This will require extensive knowledge of the vehicle systems and training in malfunction isolation and correction. In order to provide this training as many as possible of the numerous malfunctions which can occur in even a relatively simple space vehicle must be identified and simulated. Considerable effort has been devoted to this area in the Mercury training and development program and it should become an increasingly important feature of future programs.

The physical conditions (heat, acceleration, and so forth) associated with space flight are simulated to permit the trainees to adapt to these stressors in order that during the actual flight such stimuli may be less disturbing. Present measures of the adaptation process are inadequate to provide criteria for training progress. A second purpose for the familiarization program was to give the trainees an opportunity to learn the specific skills required to minimize the effects of these factors on their performance. However, in many cases, the skills required have not been fully identified or validated. For example, in developing straining techniques for meeting increased acceleration, the efficacy of a straining technique has not been fully demonstrated nor has the technique itself been adequately described. As yet, inadequate data are available on the effects of combining physical stress factors. Therefore, it is difficult to determine the extent to which the increased cost and difficulty of providing multiple stress simulation is warranted. In the present program, it has been possible to simulate both reduced atmospheric pressure and acceleration on the centrifuge. Initial experience seems to indicate that this is desirable but not critical. However, further data on the interacting effects of these stresses are required before any final conclusions can be developed.

A factor in space flight not yet adequately simulated for training purposes is weightlessness. Short periods of weightlessness have been used in the present program, as described previously. True weightlessness can be achieved through too short a period to be fully adequate for training purposes. On the other hand, ground simulation methods using water seem to be too cumbersome and unrealistic to be fully acceptable substitutes. At the present time, this lack of adequate simulation does not seem to be critical since the effects of weightlessness on performance appear to be minor and transitory. Should early space flights uncover more significant problems, greater efforts will be justified in developing weightless simulation methods.

Finally, it seems important to reiterate the requirements for re-producing adequate motivational conditions in the training program. The basic task of the Astronaut is to make critical decisions under adverse conditions. The results of the decisions he makes involve not just minor discomforts or annoyances, but major loss of equipment and even survival. Performance of this task requires a vigilance and decision-making capability difficult to achieve under the artificial conditions of ground simulation. It appears probable that training in ground devices should be augmented with flight operations to provide realistic operational conditions.

Each of the Mercury flights would carry only one man. This memorandum outlines the process for choosing the three astronauts who would be the candidates for the first two Mercury-Redstone suborbital flights.

Abe Silverstein (Director of Space Flight Programs, NASA), Memorandum for Administrator (NASA), "Astronaut Selection Procedure for Initial Mercury-Redstone Flights," December 14, 1960

* * *

3.   Since it is impractical to train all 7 Astronauts on the proper Procedures Trainer configuration, three men will be chosen as

possible pilots and all three will begin working with the capsule 7 configuration. It is hoped that the identity of the three men can be kept secure from the press. The first pilot and his alternate will not be selected until approximately one week before the launch date of the first manned Redstone. The identity of the two-man flight crew for this flight would thus not be available for announcement to either the Astronauts or the press until approximately a week before the flight. . . .

4. An astronaut Flight Readiness Board consisting of five men will be established with Robert Gilruth serving as chairman. Individuals on this board will evaluate the following pertinent areas of Astronaut performance.
    A. Medical
        a. General health
        b. Reaction to physical stress
        c. Weight control
    B. Technical
        a. Proficiency on capsule attitude control simulators
        b. Knowledge of capsule systems
        c. Knowledge of mission procedures
        d. Ability to contribute to vehicle design and flight procedures
        e. General aircraft flight experience
        f. Engineering and scientific background
        g. Ability to observe and report flight results
    C. Psychological
        a. Maturity
        b. Motivation
        c. Ability to work with others
        d. Ability to represent Project Mercury to public
        e. Performance under stress

Expert witnesses in the various areas will be called before the Board. Based on evaluations by the Board, the actual selection of the three men and of the final two-man flight crew will be made by the Director of Project Mercury.

5. The Astronauts themselves will be asked to submit to the Board chairman their recommendations for the three

best-qualified pilots, excluding themselves. This input will be known by the chairman only and will be used as an additional factor in the selection.
6. The Flight Readiness Board will meet either during the week of December 26 or January 2.

The suborbital Mercury-Redstone (MR)-3 mission scheduled for March 1961 was planned as the first flight with an astronaut aboard. But after the chimpanzee-carrying MR-2 deviations from plan, NASA decided to postpone MR-3 and insert another test flight into the schedule to make sure that the problem with the Redstone booster was fixed. That flight, launched on March 24, was a success. If not for the booster problem, it is possible that the MR-3 mission could have sent an American into space, if not into orbit, before Soviet cosmonaut Yuri Gagarin carried out his one-orbit flight on April 12, 1961. If the United States had been first to send a human into space, subsequent history might well have been quite different.

With the success of the March test flight, the way was cleared for the first U.S. astronaut flight. Press and public interest in the event was growing to the point that President Kennedy's science adviser Jerome Wiesner raised with the new president and his national security adviser McGeorge Bundy the wisdom of providing live television coverage of the launch. Wiesner wanted to be sure that the top policy levels of the White House, including the president, were fully informed regarding nontechnical aspects of the flight. Wiesner was also critical of the 1958 decision by President Eisenhower to limit the original group of astronauts to military pilots; this policy was reversed in 1962 as NASA selected its second group of astronauts, which included civilian pilot Neil Armstrong.

Concern over the wisdom of televising the flight grew after the Gagarin flight; the possibility of a televised failure soon after the Soviet success was troubling to the White House. Ultimately it was Kennedy himself who made the decision that the chances of a launch failure were low enough to go ahead with televising the launch.

Jerome B. Wiesner, Memorandum for Dr. [McGeorge] Bundy, "Some Aspects of Project Mercury," March 9, 1961

We have an ad hoc panel which is making a technical review of the National Aeronautics and Space Administration project to put a man in an earth orbit, Project Mercury. The time is now nearing when man will be first introduced into the system in a sub-orbital launch using a modified Redstone booster. Although the interest of the panel is primarily in the technical details, two phases of the operation which fall mostly outside the technical area have caused them considerable questioning, and I would like to take this opportunity to bring them to your attention.

1.   Many persons involved in the project have expressed anxiety over the mounting pressures of the press and TV for on-the-spot coverage of the first manned launch. Our panel is very concerned that every precaution should be taken to prevent this operation from becoming a Hollywood production, because it can jeopardize the success of the entire mission. The people in the blockhouse and in the control center are not professional actors, but are technically trained people involved in a very complex and highly coordinated operation. The effect of TV cameras staring down their throats during this period of extreme tension, whether taped or live, could have a catastrophic effect. Similarly, following a manned launch and recovery, the astronaut must be held in a confined area for a considerable time period so that the doctors can accomplish the debriefing which will produce the basic information on possible effects of space flight on man. The pressures from the press during this time period will probably be staggering, but should be met with firmness . . .

Our panel does not profess to be expert in the field of public relations, but the overriding need for the safety of the astronaut and the importance to our nation of a successful mission make them feel that the technical operation should have first consideration in this program. The sub-orbital launch will, in fact, be man's first venture into space. It is enough different from the X15 program to require special consideration. It is my personal opinion that in the imagination of many, it will be viewed in the same category as Columbus' discovery of the new world. Thus, it is an extremely important venture and should be exploited properly by the Administration.

2. Some members of the panel (and other individuals who have contacted me privately) believe that the decision by the previous Administration, that the astronauts should be military personnel, was wrong. They point out that NASA was created expressly for the purpose of conducting peaceful space missions, and the orbiting of a military astronaut will be identified by the world in general as a military gesture, and is sure to be seized upon by the U.S.S.R. for propaganda purposes.

My personal feeling is that any change in status (such as asking the astronauts to become civilians) at this late date will be recognized for what it is, an artificial maneuver. Nevertheless, it might be desirable for this Administration to review the past decision and perhaps lay plans by which astronauts selected for later manned space programs could be given the option to become civilians . . .

Another of Wiesner's concerns was whether NASA had adequately considered all aspects of flight readiness, and in particular the likely impacts of space travel on the astronaut. Wiesner asked the President's Science Advisory Committee to take an independent look at the issue. An ad hoc group was formed to carry out that assessment and reported that, while "the area of greatest concern to us has been the medical problem of the pilot's response to the extreme physical and emotional strains which space flights will involve," overall spaceflight was "a high risk undertaking but not higher than we are accustomed to taking in other ventures." As the panel was completing its work, NASA heard rumors that because of high uncertainty about the impacts of a spaceflight on a human, the panel would recommend up to fifty more chimpanzee flights before launching an astronaut into space. On hearing such a rumor, Robert Gilruth of the Space Task Group was reported to have commented, "We might as well move the program to Africa." While the panel's report did not include such a recommendation, it did express significant reservations about the limited data on likely impacts of spaceflight on a human. But the same day that the panel submitted its report, the Soviet Union launched Yuri Gagarin into orbit and successfully returned him to Earth. The Soviet success provided at least an initial demonstration that a human could fly in space and return to Earth in good health.

President's Science Advisory Committee, "Report of the Ad Hoc Mercury Panel," April 12, 1961

***

VII. CONCLUSIONS

1. The [Mercury] program is a reasonable step in attaining manned space flight. It represents the highest degree of technical advancement available at the time of its inception.

2. The system is a complicated one and is made so largely by the automatic devices, which are often duplicated, plus the alternate manual control and safety devices.

3. The system is not completely reliable and cannot be made so in the foreseeable future. It is not more unreliable then could have been predicted at its inception. The thought and organization which have gone into making it as reliable as possible have been careful and thorough and most of the problems have been thought through. There does not appear to be any shortage of funds for reliability and safety measures.

4. Manned Mercury flights will definitely be a hazardous undertaking, although related to such initial efforts as the flights of the Wright Brothers, Lindbergh flight, and the X-series of research aircraft.

5. A suborbital flight or flights are needed as a prelude to orbital flight. They will check out the pilot's performance, including his ability to orient the capsule in flight, adding elements which cannot be adequately simulated such as the anxiety and stress of a real flight and the extension of weightlessness to a five minute period, all under conditions where the risk is very much less than in orbital flight since descent in a reasonably accessible recovery area is assured under all conditions.

6. The presence of a man in the capsule will very greatly increase the probability of a successful completion of the mission over uninhabited or primate flights. One of the possible conclusions of the Mercury program is that the design philosophy of future vehicles may change from using the man as backup to the automatic system to designing automatic mechanisms as a backup to the man.

7. We urge that NASA appoint a group of consultants to plan and implement a full-scale crash effort on the Johnsville centrifuge and at other appropriate laboratories to obtain essential measurements under as many kinds of combined stresses as possible. The measurements should be sufficient to permit correlations between man and primates with enough certainty to estimate the human margin of reserve during the anticipated stresses of space flight. Substantial data should be on hand prior to committing an astronaut to the first Mercury flight. In view of the limited time available, and commitments of the Space Task Group Medical personnel to MR-3, we urge that additional qualified personnel be recruited to accomplish the studies.

8. We recommend consideration of including a chimpanzee in the forthcoming MA-3 flight. This is designed for an abort of the McDonnell capsule, complete with life support systems, from an Atlas booster just prior to capsule insertion into orbit.

9. We urge a considerable expansion of the scientific base of the medical program. Working consultants, additional in-house personnel and sufficient funds to permit implementation of a sound program, based on the resources and capabilities of several university laboratories and utilizing additional contracts with DOD and other government facilities, are essential if we are to insure reasonable programs toward orbital flight.

GENERAL CONCLUSION

The Mercury program has apparently been carried through with great care and there is every evidence that reasonable steps have been taken to obtain high reliability and provide adequate alternatives for the astronaut in the event of an emergency. Nevertheless, one is left with the impression that we are approaching manned orbital flight on the shortest possible time scale so that the number of over-all system tests will necessarily be small. Consequently, although it is generally assumed by the public that manned flight will not be attempted until we are "certain" to be able to return the man safely and that we are more conservative in our attitude toward human life than is the USSR, the fact seems to be that manned flight will inevitably involve a high degree of risk and that the USSR will have carried out a more extensive preliminary program particularly in animal studies than we will before sending a man aloft.

It is difficult to attach a number to the reliability. The checkout procedures on individual components and for the flight itself are meticulous. There appear to be sufficient alternative means by which the pilot can help himself if the already redundant mechanical system fails. However, there is no reliable current statistical failure analysis and although we feel strongly that such analyses should certainly be brought up to date before the first orbital flight we see no likelihood of obtaining an analysis which we would really trust. One can only say that almost everything possible to assure the pilot's survival seems to have been done.

The area of greatest concern to us has been the medical problem of the pilot's response to the extreme physical and emotional strains which space flights will involve. On this score the pilot training has been thorough and it has been demonstrated that a man can perform under the conditions of acceleration and weightlessness to which he will be subjected. Nevertheless, the background of medical experimentation and test seems very thin. The number of animals that will have undergone flights will be much smaller than in the USSR program. Consequently, we are not as sure as we would like to be that a man will continue to function properly in orbital missions although the dangers seem far less pronounced in a suborbital flight.

Altogether, the probability of a successful suborbital Redstone flight is around 75 percent. The probability that the pilot will survive appears to be around 90 to 95 per cent although the NASA estimates are somewhat higher. This does not appear to be an unreasonable risk, providing the known problems are taken care of before the flight, and those of our members who have been very close to the testing of new aircraft fell that the risks are comparable to those taken by a test pilot with a new high performance airplane.

It is too early to say anything as definite for the risks of orbital flight. Nevertheless, if the planned program of tests is carried through it seems probable that the situation at the time of the first flight will be comparable to that for a Redstone flight now— a high risk undertaking but not higher than we are accustomed to taking in other ventures.

The astronaut chosen to make the first U.S. spaceflight was Alan Shepard, who, along with John Glenn, was an acknowledged leader of the Mercury 7. After several delays due to weather and equipment problems, Shepard was launched at 9:34 a.m. on May 5, 1961, on a 15-minute, 22-second suborbital trajectory that he described as "a pleasant ride." Shepard's Freedom 7 *spacecraft reached an altitude of 116.5 miles and landed 301 miles downrange from his launchpad at Cape Canaveral. The flight had minimal technical problems and Shepard returned in excellent physical condition.*

Shepard and the other six Mercury astronauts were welcomed to the White House by President Kennedy on May 8. After the White House meeting, Shepard paraded down Pennsylvania Avenue to an enthusiastic welcome by a large, hastily assembled crowd. On that same day Kennedy received the recommendation that he set human landings on the Moon as a national goal (See chapter 3). It is worth speculating whether Kennedy would have so readily accepted that recommendation if Shepard's flight had not been such a technical and publicly applauded success.

In the aftermath of his flight, there were a variety of suggestions that Shepard, now a national hero, take on a public role as an advocate of the accelerated space program proposed by President Kennedy in his May 25, 1961, speech to a joint session of Congress. One such suggestion came from James Hagerty, president of the ABC television network. Hagerty had served as President Eisenhower's press secretary between 1953 and 1961, and in that capacity had dealt often with the media issues brought to the fore by Soviet "space spectaculars." Upon his departure from Washington with the end of the Eisenhower administration, he keenly understood the excitement of spaceflight and in that context tried, as this letter suggests, to play on the public's interest in the astronauts to aid his new organization, ABC, by organizing a joint appearance by the first two humans in space, Yuri Gagarin and Alan Shepard. The new NASA administrator, James Webb, a conservative Democrat with extensive Washington experience, including tenure as deputy secretary of state and director of the Bureau of the Budget, reacted negatively to Hagerty's proposal, and it was not pursued.

James E. Webb (Administrator, NASA), to James C. Hagerty (Vice President, American Broadcasting Company), June 1, 1961

Dear Jim:

Since our discussion on May 19th I have given a good deal of thought to the proposal outlined in the letter you delivered to me on that date, that Alan Shepard and Yuri Gagarin appear together in New York City on a nation-wide telecast. I cannot see how Shepard's appearance would serve a useful purpose, and I believe it could be detrimental to the best interests of the United States.

Although, as I told you, your proposal involves national policy questions beyond my own direct responsibility, I feel that it is my duty to state my conviction that the whole plan is unwise.

The Mercury flight of Alan Shepard was performed before the eyes of the whole world. He reported his immediate experience and reactions at the press conference on May 8th in Washington. On June 6th Shepard and other members of the Space Task Group will give a full report on the results of the flight at a scientific and technical conference in Washington which will be widely reported and whose proceedings will be published. Further reporting could add nothing significant.

The free and open way in which we have proceeded to share our manned-flight experience with the world is in marked contrast to Soviet secrecy and their unsupported and conflicting descriptions of the Gagarin flight.

If, as you have proposed, Gagarin would be free to tell his story in whatever manner he so desired, it is fair to assume it would not be in a full and complete factual framework but rather in the same framework as previous reports.

Why then should we permit the Soviet Union to blunt the

impact of the open conduct of our program by the use of a nation-wide telecast as a propaganda forum?

From past experience, the Russians might very well use Gagarin's appearance here in the United States to announce and to exploit, again without full facts, and to a large audience, another Russian manned flight, timed to coincide with his appearance here—perhaps a flight of two or three persons. In such a situation, to compare Shepard's suborbital flight with that of Gagarin, or with some other Russian achievement, would be inconsistent with the reporting of the flight as only one step in the U.S. ten-year program for space exploration . . .

The second flight of the Mercury program, astronaut Gus Grissom's suborbital MR-4 mission on July 21, 1961, proved somewhat less successful than Alan Shepard's May 5 flight because of the loss of the capsule in the ocean. On Grissom's mission, an explosively actuated system was to be used to blow open the seventy titanium bolts that secured the spacecraft's side hatch to the doorsill. During the water recovery effort, there was a premature explosion, and the hatch blew off. Grissom vacated the spacecraft immediately and, after nearly drowning because of water getting into his space suit, was retrieved after being in the water for about four minutes. This debriefing, which took place shortly after the incident, captures Grissom's near real-time description of what happened.

MR-4 Technical Debriefing Team, Memorandum for Associate Director (NASA), "MR-4 Postflight Debriefing of Virgil I. Grissom," July 21, 1961, with attached "Debriefing"

* * *

11. Recovery—On landing, the capsule went pretty well under the water. Out the window, I could see nothing but water and it was apparent to me that I was laying pretty well over on my left side and little bit head down. I reached the rescue aids switch and I heard the reserve chute jettison and I could see the canister in the water through the periscope. Then, the capsule righted itself rather rapidly and it was apparent to me that I was in real good shape, and I reported this. Then I got

ready to egress. I disconnected the helmet from the suit and put the neck dam up. The neck dam maybe had been rolled up too long, because it didn't unroll well. It never did unroll fully. I was a little concerned about this in the water because I was afraid I was shipping a lot of water through it. In fact, the suit was quite wet inside, so I think I was. At this point, I thought I was in good shape. So, I decided to record all the switch positions just like we had planned. I took the survival knife out of the door and put it into the raft. All switches were left just the way they were at impact, with the exception of the rescue aids and I recorded these by marking them down on the switch chart in the map case and then put it back in the map case. I told Hunt Club they were clear to come in and pick me up whenever they could. Then, I told them as soon as they had me hooked and were ready, I would disconnect my helmet, take it off, power down the capsule, blow the hatch, and come out. They said, "Roger," and so, in the meantime, I took the pins off both the top and the bottom of the hatch to make sure the wires wouldn't be in the way, and then took the cover off the detonator.

12. I was just waiting for their call when all at once, the hatch went. I had the cap off and the safety pin out, but I don't think that I hit the button. The capsule was rocking around a little but there weren't any loose items in the capsule, so I don't see how I could have hit it, but possibly I did. I had my helmet unbuttoned and it wasn't a loud report. There wasn't any doubt in my mind as to what had happened. I looked out and saw nothing but blue sky and water starting to ship into the capsule. My first thought was to get out, and I did. As I got out, I saw the chopper was having trouble hooking onto the capsule. He was frantically fishing for the recovery loop. The recovery compartment was just out of the water at this time and I swam over to help him get his hook through the loop. I made sure I wasn't tangled anyplace in the capsule before swimming toward the capsule. Just as I reached the capsule, he hooked it and started lifting the capsule clear. He hauled the capsule away from me a little bit and didn't drop the horsecollar down. I was floating, shipping water all the time, swallowing some, and I thought

one of the other helicopters would come in and get me. I guess I wasn't in the water very long but it seemed like an eternity to me. Then, when they did bring the other copter in, they had a rough time getting the horsecollar to me. They got in within about 20 feet and couldn't seem to get it any closer. When I got the horsecollar, I had a hard time getting it on, but I finally got into it. By this time, I was getting a little tired. Swimming in the suit is difficult, even though it does help keep you somewhat afloat. A few waves were breaking over my head and I was swallowing some water. They pulled me up inside and then told me they had lost the capsule.

***

NASA canceled the final two planned suborbital Mercury flights as no longer necessary. The Shepard and Grissom missions had demonstrated that the Mercury capsule was "spaceworthy" and that an astronaut could carry out various tasks after launch and while in a weightlessness state. That ability would be tested as Project Mercury approached its first orbital flight.

NASA had to address the question of what astronauts might say while in orbit. During the flight of Yuri Gagarin in April 1961 and a second, seventeen-orbit flight of Gherman Titov in August 1961, the Soviet cosmonauts had delivered propaganda messages in support of Soviet interests. The question was whether U.S. astronauts should follow that precedent. Recognizing that their orbital flights would take them over many nations, NASA prepared guidelines for what the astronauts might say as they flew above particular countries.

These guidelines reflected the importance of the Mercury program to the larger Cold War rivalry with the Soviet Union. One of the key aspects of the early space race involved convincing the peoples of nonaligned countries of the superiority of the United States and its way of life over the Soviet Union. Directly speaking to some of these people from space might help sway their opinion. At the same time, there was a desire to appear genuine, unscripted, and nonpropagandistic. NASA eventually decided that U.S. astronauts should eschew propaganda messages and stick to bland commentary. The four Mercury astronauts who flew orbital missions, John Glenn, Scott Carpenter, Wally

Schirra, and Gordon Cooper, performed their roles quite well, making noncontroversial remarks that were heard around the world.

Dr. Robert B. Voas (Training Officer, NASA), Memorandum for Astronauts, "Statements for Foreign Countries During Orbital Flights," November 7, 1961

1. The undersigned has attempted to get guidance within the NASA organization on the policy to be pursued in making statements of possible political significance from the Mercury capsule. In pursuing this question, he was referred to Mr. Goodwin of NASA Headquarters. Mr. Goodwin made the following suggestions. These he apparently discussed with Mr. Lloyd [NASA head of public affairs] and the Administrator and they have their approval.

   (a) It is essential that any statements made by the Astronauts appear to be spontaneous, personal and unrehearsed. He felt that there was a general agreement that statements made by the Russian Cosmonauts were not effective and backfired. There was a general feeling that they were being used inappropriately for propaganda. He agreed strongly with our own feeling that any political statement would look out of place. Mr. Goodwin also thought that statements in a foreign language could be dangerous, because unless there was a good basis to believe they were spontaneous, they would appear to be contrived. Thus, if the Astronaut spoke in Hindustani during the flight, the inevitable question could be raised in the press conference following the flight, "How did the Astronaut come to know Hindustani?" Unless he could show that it was a course given in the high school or college which he attended, it would be obvious that this statement had been politically inspired. The one point at which a foreign language might effectively be used would be over the Mexican station. Here, a few words of Spanish, such as, "Saludos Amigos," might be quite appropriate and since simple Spanish phrases are known by many Americans, it would not appear contrived.

2. While Mr. Goodwin did not feel that either a political state-
    ment as such, or statements in foreign languages, would be
    useful, he did fell that descriptions by the Astronaut in En-
    glish of the terrain over which he was passing and personal
    statements of how he felt and reacted to the situation would
    be highly desirable and effective if released to foreign per-
    sonnel. The primary requirement here on the Astronaut
    would be to be familiar enough with the political boundar-
    ies to be able to relate his observation of the ground to the
    countries over which he is passing. This way, he could re-
    port, for example, "I see it is a sunny day in Nigeria," or "I
    can still see Zanzibar, but it looks like rain is on the way."
    To these observations related directly to the country should
    be added any personal observations such as, "I feel fine;
    weightlessness doesn't bother me a bit; it's just like flying in
    an aircraft, etc." In all such statements, care must be taken
    not to make them appear to be contrived, maudlin or too
    effusive. Rather, they should be genuine, personal and with
    immediate impact. Mr. Goodwin points out that the ideas
    and words expressed are more important to communica-
    tion than using the actual language of the country. If the
    experience which the Astronaut is having can be expressed
    in personal, simple, meaningful terms, when translated, this
    will be far more effective than a few words in a foreign lan-
    guage which, in the long run, might appear contrived.

***

*For orbital flights, the Mercury spacecraft would be capable of
supporting a human in space for as long as three days. As a
launch vehicle for the orbital missions, NASA chose the more
powerful Atlas booster, a modified intercontinental ballistic missile. On November 29, 1961, a test flight of the Mercury/Atlas
combination took place, with the chimpanzee Enos occupying
the capsule for a two-orbit ride before being successfully recovered in an Atlantic Ocean landing. The way was clear for the first
American to go into orbit. That person would be John Glenn.*

*Glenn's flight, originally scheduled for December 1961, took
place after several delays. Not until February 20, 1962, did
NASA launch Glenn. He became the first American to circle
the Earth, making three orbits in his* Friendship 7 Mercury

spacecraft. Glenn experienced a potentially disastrous event when a landing bag on the back side of his spacecraft, programmed to inflate a few seconds before splashdown to help cushion the impact, possibly inflated in orbit. The landing bag was located just inside the heat shield, an ablative material meant to burn off during reentry, and was held in place in part by a retropack of three rocket motors that would slow the capsule down and drop it from orbit. Because of this apparent problem, Glenn had to return to Earth after only three orbits, instead of a planned seven, and had to leave the retropack in place during his fiery reentry, hoping that it would hold the heat shield in place. It did, and Glenn returned safely.

Glenn's flight brought a welcome increase in national pride. The public embraced Glenn as a personification of heroism and dignity. Glenn addressed a joint session of Congress and participated in several ticker-tape parades around the country.

In this debriefing shortly after his recovery, Glenn describes what took place during his flight.

John Glenn, "Brief Summary of MA-6 Orbital Flight," February 20, 1962

There are many things that are so impressive, it's almost impossible to try and describe the sensations that I had during the flight. I think the thing that stands out more particularly than anything else right at the moment is the fireball during the reentry. I left the shutters open specifically so I could watch it. It got a brilliant orange color; it was never too blinding. The retropack was still aboard and shortly after reentry began, it started to break up in big chunks. One of the straps came off and came around across the window. There were large flaming pieces of the retropack—I assume that's what they were—that broke off and came tumbling around the sides of the capsule. I could see them going on back behind me then making little smoke trails. I could also see a long trail of what probably was ablation material ending in a small bright spot similar to that in the pictures out of the window taken during the MA-5 flight. I saw the same spot back there and I could see it move back and forth as the capsule oscillated slightly. Yes, I think the reentry was probably the most impressive part of the flight.

Starting back with highlights of the flight: Insertion was

normal this morning except for the delays that were occasioned by hatch-bolt trouble and by the microphone fitting breaking off in my helmet. The weather cleared up nicely and after only moderate delays, we got off.

Lift-off was just about as I had expected. There was some vibration. Coming up off the pad, the roll programming was very noticeable as the spacecraft swung around to the proper azimuth. There also was no doubt about when the pitch programming started. There was some vibration at lift-off from the pad. It smoothed out just moderately; never did get to very smooth flight until we were through the high q area. At this time—I would guess a minute and fifteen to twenty seconds—it was very noticeable. After this, it really smoothed out and by a minute and a half, or about the time cabin pressure sealed off, it was smooth as could be.

The staging was normal, though I had expected a more sharp cutoff. It felt as though the g ramped down for maybe half a second. For some reason, it was not as abrupt as I had anticipated it might be . . . As the booster and capsule pitched over and the tower jettisoned, I had a first glimpse of the horizon; it was a beautiful sight, looking eastward across the Atlantic.

. . . As we were completing the turnaround, I glanced out of the window and the booster was right there in front of me. It looked as though it wasn't more than a hundred yards away. The small end of the booster was pointing toward the northeast and I saw it a number of times from then on for about the next seven or eight minutes as it slowly went below my altitude and moved farther way. That was very impressive.

I think I was really surprised at the ease with which the controls check went. It was almost just like making the controls check on the Procedures Trainer that we've done so many times. The control check went off like clockwork; there was no problem at all. Everything damped when it should damp and control was very easy. Zero-g was noticeable . . . I had a very slight sensation of tumbling forward head-over-heels. It was very slight; not as pronounced an effect as we experience on the centrifuge. During turnaround, I had no sensation of angular acceleration. I acclimated to weightlessness in just a matter of seconds; it was very surprising. I was reaching for switches and doing things and having no problem. I didn't at any time notice any tendency to overshoot a switch. It seemed it's just natural to acclimate to this new condition. It was very comfortable. Under the weightless condition, the head seemed to be a little farther out of the couch which made it a little easier to see [out] the window, though I could not get up quite as near to the window as I thought I might.

The rest of the first orbit went pretty much as planned, with reports to the stations coming up on schedule. I was a little behind at a couple of points but most of the things were going right according to schedule, including remaining on the automatic control system for optimum radar and communications tracking. Sunset from this altitude is tremendous. I had never seen anything like this and it was truly a beautiful, beautiful sight. The speed at which the sun goes down is very remarkable, of course. The brilliant orange and blue layers spread out probably 45 to 60 degrees each side of the sun tapering very slowly toward the horizon. I could not pick up any appreciable Zodiacal light. I looked for it closely; I think perhaps I was not enough night adapted to see it. Sunrise, I picked up in the periscope. At every sunrise, I saw little specks, brilliant specks, floating around outside the capsule. I have no idea what they were. On the third orbit, I turned around at sunrise so that I could face into the sun and see if they were still heading in the same direction and they were. But I noticed them every sunrise and tried to get pictures of them.

Just as I came over Mexico at the end of the first orbit, I had my first indication of the ASCS [Automatic Stabilization Control System] problem that was to stick with me for the rest of the flight. It started out with the yaw rate going off at about one and one-half degrees per second to the right. The capsule would not stay in orbit mode, but would go out of limits . . . I took over manually at that point and from then on, through the rest of the flight, this was my main concern. I tried to pick up the flight plan again at a few points and I accomplished a few more things on it, but I'm afraid most of the flight time beyond that point was taken up with checking the various modes of the ASCS. I did have full control in fly-by-wire and later on during the flight, the yaw problem switched from left to right. It acted exactly the same, except it would drift off to the left instead of the right. It appeared also that any time I was on manual control and would be drifting away from the regular orbit attitude for any appreciable period of time that the attitude indications would then cut off when I came back to orbit attitude.

* * *

Retrorockets were fired right on schedule just off California and it was surprising coming out of the zero-g field that retrorockets firing felt as though I were accelerating in the other direction back toward Hawaii. However, after retrofire was completed when I could glance out the window again, it was easy to tell, of course, which way I was going, even though my sensations during retrofire had been that I was going in the other direction.

* * *

Following retrofire, a decision was made to have me reenter with the retropackage still on because of the uncertainty as to whether the landing bag had been extended. I don't know all the reasons yet for that particular decision, but I assume that it had been pretty well thought out and it obviously was. I punched up .05g manually at a little after the time it was given to me. I was actually in a small g-field at the time I pushed up .05g and it went green and I began to get noise, or what sounded like small things brushing against the capsule. I began to get this very shortly after .05g and this noise kept increasing. Well before we got into the real heavy fireball area, one strap swung around and hung down over the window. There was some smoke. I don't know whether the bolt fired at the center of the pack or what happened. The capsule kept on its course. I didn't get too far off the reentry attitude. I went to manual control for reentry after the retros fired and had no trouble controlling reentry attitude through the high-g area. Communications blackout started a little bit before the fireball. The fireball was very intense. I left the shutters open the whole time and observed it and it got to be a very bright orange color. There were large, flaming pieces of what I assume was the retropackage breaking off and going back behind the capsule. This was of some concern, because I wasn't sure of what it was. I had visions of them possibly being chunks of heat shield breaking off, but it turned out it was not that.

The oscillations that built up after peak-g were more than I could control with the manual system. I was damping okay and it just plain overpowered me and I could not do anymore about it. I switched to Aux. Damp as soon as I could raise my arm up after the g-pulse to help damp and this did help some. However, even on Aux. Damp, the capsule was swinging back and forth very rapidly and the oscillations were divergent as we descended to about 35,000 feet. At this point, I elected to try to put the drogue out manually, even though it was high, because I was afraid we were going to get over to such an attitude that the capsule might actually be going small end down during part of the flight if the oscillations kept going the way they were. And just as I was reaching up to pull out the drogue on manual, it came out by itself. The drogue did straighten the capsule out in good shape. I believe the altitude was somewhere between 30,000 and 35,000 at that point.

I came on down; the snorkels, I believe, came out at about 16,000 or 17,000. The periscope came out. There was so much smoke and dirt on the windshield that it was somewhat difficult to see. Every time I came around to the sun—for I had established my roll rate on manual—it was virtually impossible to see anything out through the window.

The capsule was very stable when the antenna section jettisoned. I could see the whole recovery system just lined up in one big line as it came out. It unreeled and blossomed normally; all the panels and visors looked good. I was going through my landing checkoff list when the Capsule Communicator called to remind me to deploy the landing bag. I flipped the switch to auto immediately and the green light came on and I felt the bag release. I was able to see the water coming towards me in the periscope. I was able to estimate very closely when I would hit the water. The impact bag was a heavier shock than I had expected, but it did not bother me.

***

In summary, my condition is excellent. I am in good shape; no problems at all. The ASCS problems were the biggest I encountered on the flight. Weightlessness was no problem. I think the fact that I could take over and show that a pilot can control the capsule manually, using different control modes, satisfied me most. The greatest dissatisfaction I think I feel was the fact that I did not get to accomplish all the other things that I wanted to do. The ASCS problem overrode everything else.

An issue that received significant attention in the 1961–62 period was the absence of any women in the NASA astronaut corps. There were, of course, women working in various capacities in support of Project Mercury and other NASA efforts. In the era before digital computers, women performed many of the calculations required to determine spacecraft orbits and other parameters. Indeed, the job title for such a job was "computer." Some of these women were African Americans based at the Space Task Group in Virginia; they later became the focus of a book and then movie titled Hidden Figures.[4]

NASA in 1958 had made a conscious decision to limit applications for Project Mercury astronauts to men, reinforced by President Eisenhower's subsequent decision to limit the applicant pool to military test pilots. But the person who had overseen many of the medical tests to which astronaut candidates had been subjected, Dr. Randolph Lovelace, was curious how women would fare if they took the same tests and examinations. In 1960–61 he initiated a "Women in Space" effort and recruited nineteen women to undergo similar scrutiny; of that number thirteen passed. One of them, Jerrie Cobb, was a high-profile, award-winning pilot and flight instructor. Cobb gained widespread publicity as an advocate for including women in the next group of astronaut applicants, and in 1961 NASA administrator James Webb appointed her as a NASA consultant. Cobb was joined in her advocacy by the wife of Michigan senator Philip Hart, Janey Hart, who had also passed the medical tests. Together they lobbied NASA, the White House, and Congress in support of opening up the astronaut corps to women.

On March 15, 1962, just a few weeks after John Glenn's orbital flight, Cobb and Hart arranged through Lyndon Johnson's press secretary Liz Carpenter to meet with the vice president. Carpenter drafted a cautious letter to James Webb for Johnson to sign during the meeting. But it turned out that Johnson did not support the concept of women astronauts. He did not sign the letter during the meeting; instead, he wrote on the letter in bold script, "Let's stop this now!" and "File." The letter remained undiscovered in Johnson's files for many years, and NASA did not select its first female astronauts until 1978.

Draft Letter from Vice President Lyndon B. Johnson to NASA Administrator James Webb, March 15, 1962

THE VICE PRESIDENT
WASHINGTON
March 15, 1962

Dear Jim:

I have conferred with Mrs. Philip Hart and Miss Jerrie Cobb concerning their effort to get women utilized as astronauts. I'm sure you agree that sex should not be a reason for disqualifying a candidate for orbital flight.

Could you advise me whether NASA has disqualified anyone because of being a woman?

As I understand it, two principal requirements for orbital flight at this stage are: 1) that the individual be experienced at high speed military test flying; and 2) that the individual have an engineering background enabling him to take over controls in the event it became necessary.

Would you advise me whether there are any women who meet these qualifications?

If not, could you estimate for me the time when orbital flight will have become sufficiently safe that these two requirements are no longer necessary and a larger number of individuals may qualify?

I know we both are grateful for the desire to serve on the part of these women, and look forward to the time when they can.

Sincerely,

Lets stop this now!

Lyndon B. Johnson

File

Mr. James E. Webb
Administrator
National Aeronautics and Space Administration
Washington, D.C.

Another issue with respect to the NASA astronaut corps was the activities of the astronauts outside of their official duties. The public, of course, relished as much information as could be obtained about the Mercury 7, and NASA had facilitated the sale by the astronauts, at a considerable fee, of their personal stories to Life magazine both as a means of satisfying that thirst and as a form of financial insurance for the astronauts' families should an astronaut lose his life in spaceflight. As they became aware of this arrangement, President John F. Kennedy and his advisers questioned whether it was appropriate. The astronauts were government employees, and most government workers were prohibited from taking additional compensation for their official duties.

After his flight, John Glenn had become friendly with President Kennedy and his brother Robert. On a visit to the Kennedy vacation compound in Hyannis Port, Massachusetts, in summer 1962 Glenn had discussed with the two Kennedys the specifics of the Life *contract and the fact that it covered the additional expenses linked to the astronauts and their families constantly being in the public eye. In an August 1962 White House discussion, Kennedy gave his permission for extending a revised version of the* Life *contract to the second group of astronauts that NASA was then recruiting. NASA worked to refine its policies but never found a fully satisfactory solution that balanced the rights, needs, and privileges of the astronauts with federal regulations regarding private activities of government employees.*

Richard L. Callaghan (NASA), Memorandum for Mr. James E. Webb, "Meeting with President Kennedy on Astronaut Affairs," August 30, 1962

* * *

The White House meeting lasted some 30 minutes. The President at the outset stated generally that he felt the astronauts should be permitted to continue to receive some money for writings of a personal nature inasmuch as they did seem to be burdened with expenses they would not incur were they not in the public eye. He felt there should be stricter control of their investments. He cited the proffer of homes in Houston as an example of the type of situation that should be avoided in the future.

* * *

Without detailing the discussion further, the following portrays my impression of the conclusions reached at the meeting with the President.

1. The President leaves to your discretion the preparation of such refinements in NASA's proposed policy revisions as are necessary to:
    a. Permit the continued sale by the astronauts of their personal stories, whether through a LIFE-type contract or otherwise.
    b. Extend the prohibition against commercial endorsements.
    c. Provide reasonable supervision of the astronauts' investments (although this need not be a specifically stated part of the policy, the astronauts are to understand that such supervision is inherent in the policy).
    d. Serve generally as a model of administration policy.
2. Within the framework of its policy NASA should attempt to:
    a. Make available to all news media at debriefings and press conferences a more comprehensive presentation of the official aspects of space missions in which the astronauts participate.
    b. Afford to the press additional access to NASA personnel (including the astronauts), NASA installations, and NASA facilities to the extent that such access does not impede the agency's programs or activities.
    c. Edit more stringently the material made available by the astronauts for publication.
    d. Restrict extravagant claims by publishers who attempt to overemphasize the exclusive nature of material received from the astronauts for publication.

Three more Mercury flights took place during 1962 and 1963. Scott Carpenter made a three-orbit flight on May 20, 1962. Carpenter got distracted during the flight, ignored instructions from the ground, used too much fuel in maneuvering his spacecraft, and was three seconds late in initiating his return to Earth. He landed 250 miles from his intended point, and it was some time before his safe return was confirmed. Walter Schirra flew an almost perfect six-orbit flight on October 3, 1962. The capstone of Project Mercury came on the May 15–16, 1963, flight of Gordon Cooper, who circled the Earth twenty-two times in thirty-four hours. There were a series of spacecraft malfunctions in the latter stages of Cooper's flight, and ultimately he had to take manual control for reentry. The Mercury astronauts lobbied President Kennedy for a seventh flight but were not successful in their appeal.

Christopher Columbus Kraft Jr. served as senior flight director for all the Mercury missions. Kraft had joined NACA in 1944 and became a member of the Space Task Group upon its inauguration in 1958. He was thus in an excellent position to provide an overview of the results of Project Mercury.

Christopher C. Kraft Jr., "A Review of Knowledge Acquired from the First Manned Satellite Program," undated, but 1963

SYNOPSIS

With the completion of the Mercury program, science has gained considerable new knowledge about space. In more than 52 hours of manned flight, the information brought back has changed many ideas about space flight. Design problems occupied the first and major portion of the Mercury program. The heat shield, the shape of the Mercury spacecraft, the spacecraft systems, and the recovery devices were developed. Flight operations procedures were organized and developed and a training program for both ground and flight crew was followed. Scientific experiments were planned with Man in the loop. These included photography, extra spacecraft experiments, and observation or self-performing types of experiments.

But the real knowledge of Mercury lies in the change of the basic philosophy of the program. At the beginning, the capabilities of Man were not known, so the systems had to be designed to function automatically. But with the addition of Man to the loop, this philosophy changed 180 degrees since primary success of the mission depended on Man backing up automatic equipment that could fail.

INTRODUCTION

***

The three basic aims of Project Mercury were accomplished less than five years from the start of the program. The first U.S.

manned space flight program was designed to (1) put man into Earth orbit (2) observe his reactions to the space environment and (3) bring him back to Earth safely at a point where he could be readily recovered. All of these objectives have been accomplished, and some have produced more information than we expected to receive from conducting the experiment.

The whole Mercury project may be considered an experiment, in a certain sense. We were testing the ability of a man and machine to perform in a controlled but not completely known environment.

The control, of course, came from the launch vehicle used and the spacecraft systems included in the vehicle. Although we knew the general conditions of space at Atlas insertion altitudes, we did not know how the specific environment would affect the spacecraft and the man. Such conditions as vacuum, weightlessness, heat, cold, and radiation were question marks on the number scale. There were also many extraneous unknowns which would not affect the immediate mission but would have to be considered in future flights. Such things as visibility of objects, the airglow layer, observation of ground lights and landmarks, and atmospheric drug effects were important for future reference.

***

A series of flight tests and wind tunnel tests were conducted to get the answers to some of the basic questions. First, would the ablation principle work in our application? Could we conduct heat away from the spacecraft body by melting the fiberglass and resin material? How thick would the shield have to be for our particular conditions? What temperatures would be encountered and for what time period would they exist? Early wind tunnel tests proved in theory that the saucer shaped shield would protect the rest of the spacecraft from heat damage. The flight test on the heat shield must prove the theory. In February 1961, we made a ballistic flight in which the spacecraft reentered at a sharper angle than programmed and the heat shield was subjected to great than normal heating. The test proved the heat shield material to be more than adequate.

The Mercury spacecraft did not start with the familiar bell shape. It went through a series of design changes and wind tunnel tests before the optimum shape was chosen. The blunt shape had proven best for the nose cone reentry. Its only drawback was the lack of stability. We next tried the cone-shaped spacecraft, but wind tunnel testing proved that heating on the afterbody would be too severe, although the craft was very stable in reentry. After two more trial shapes, the blunt bottom cylinder on a cone shape came into being. It was a complete cycle from the early concepts of manned spacecraft, but it was only the first of a series of changes in our way of thinking of the flight program and its elements.

A second part of design philosophy thinking came in connection with the use of aircraft equipment in a spacecraft. We had stated at the start of the program that Mercury would use as much as possible the existing technology and off-the-shelf items in the design of the manned spacecraft. But in many cases off-the-shelf equipment would just not do the job. Systems in space are exposed to conditions that do not exist for aircraft within the envelope of the atmosphere. Near absolute vacuum, weightlessness and extremes of temperatures makes equipment react differently than it does in aircraft. We had to test equipment in advance in the environment in which it was going to be used. It produced an altered concept in constructing and testing a spacecraft. Although aircraft philosophy could be adapted, in many cases, aircraft parts could not perform in a spacecraft.

The third part of the design philosophy, and perhaps the most important one in regard to future systems, is the automatic systems contained in the Mercury spacecraft. When the project started, we had no definitive information on how Man would react in the spacecraft system. To insure that we returned the spacecraft to Earth as planned, the critical functions would have to be automatic. The control system would keep the spacecraft stabilized at precisely thirty-four degrees above the horizontal. The retrorockets would be fired by an automatic sequence under a programmed or ground command. The drogue and main parachutes would deploy when a barostat inside the spacecraft indicated that the correct altitudes had been reached. The Mercury vehicle was a highly automatic system and the man essentially was riding along as a passenger, an observer. At all costs, we had to make sure that the systems worked.

But we have been able to take advantage of Man's capability in space. It started from the first manned orbital flights. When some of the thrusters became inoperative on John Glenn's flight, he was able to assume manual control of the spacecraft in order to fly the full three orbits planned in the mission. When a signal on the ground indicated the heat shield had deployed, Glenn by-passed certain parts of the retrosequence manually and retained the retropack after it had fired. In this way, he insured that the heat shield would stay in place during reentry and the spacecraft would not be destroyed by excessive heating. When oscillations built up during reentry, Glenn utilized his manual capability to provide damping using both the manual and fly-by-wire thrusters. The pilot's role in manned space flight was assuming a more important aspect.

Carpenter's flight again emphasized the ability of the pilot to control the spacecraft through the critical reentry period. Excess fuel was used in both of these orbital flights. Schirra's task was to determine if Man in the machine could conserve fuel for a long flight by turning off all systems in drifting flight. It was a task that could not be accomplished by a piece of automatic equipment in the confined area of the Mercury spacecraft. Schirra also was able to exercise another type of pilot control. It was the fine control necessary to adjust pressure suit air temperature to produce a workable environment . . . The Cooper flight was a fitting climax to the Mercury program. Not only did it yield new information for other spacecraft programs, but it demonstrated that Man had a unique capability to rescue a mission that would not have been successfully completed with the automatic equipment provided.

Man serves many purposes in the orbiting spacecraft. Not only is he an observer, he provides redundancy not obtainable by other means, he can conduct scientific experiments, and he can discover phenomenon not seen by automatic equipment.

But most important is the redundancy, the ability of another system to take over the mission if the primary system fails. Duplicate systems are designed to prevent bottlenecks in the operation of the systems. The single point failure caused the false heat shield signal in Glenn's flight. After the mission was successfully completed, we conducted an intense design review to see if there were any more of these single points in the spacecraft that needed redundancy of design for safe operation. We found many areas where the failure of one component could trigger a whole series of unfavorable reactions. This type of problem had been brought about by the design philosophy originally conceived because of the lack of knowledge of Man's capability in a space environment.

**

Our experience with the Mercury network changed our thinking about the operation of this worldwide tracking system for manned flights. In the initial design of the network, we did not have voice communication to all the remote sites.

But we soon found that in order to establish our real time requirement for evaluating unusual situations, we needed the voice link. When we started the program, the determination of the orbital ephemeris was a process that could take several orbits to establish. We could not tolerate such a condition in a manned flight so we set up a worldwide network which would maintain contact with the astronaut approximately 40 minutes out of every hour. But continuous voice contact with the astronaut has proven unnecessary and in many cases undesirable. While we retain the capability to contact an astronaut quickly, we have tried to reduce the frequency of communications with the spacecraft.

In designing and modifying a spacecraft, it is also possible to learn something more than tangible changes or hardware design. We learned about the reliability requirement and the very important need to check details carefully. It is a requirement that cannot be designed into a system on the drawing board. It actually consists in developing a conscientious contractor team that will take care to follow procedures and deliver a reliable product. Then it takes a careful recheck by the government team to insure that reliability has actually been built into the product. The smallest mistake in a man rated system can bring totally unexpected results. The unexpected is the rule in the unknown, and if Man is going to live in the region beyond our atmosphere, he is going to live under rules or not at all. We have been aware of these new rules from the start of the satellite program, but they have not been brought to our attention so vividly as they have in the manned flight program.

***

AEROMEDICAL EXPERIMENTS

While we can redesign the equipment to accomplish the mission, we cannot redesign the man who must perform in space. Aeromedical experiments for new knowledge about space must simply answer one question. Can Man adapt to an environment which violates most of the laws under which his body normally operates? The answer to the question at the end of the Mercury program seems to be an unqualified yes, at least for the period of one to two days.

The crushing acceleration of launch was the first concern. We knew he would be pressed into his couch by a force equal to many times the weight of his own body. It was not definitely known whether he would be able to perform any piloting functions under these high "g" forces. The centrifuge program was started and the astronauts tested under this stress proved that Man was not as fragile or helpless as we might have supposed. In addition to being able to withstand heavy acceleration, a method was developed of straining against the force and performing necessary pilot control maneuvers.

Weightlessness was a real aeromedical unknown and it was something that the astronauts could not really encounter on the ground. The ability to eat and drink without gravity was one serious question we had to answer. In the weightless condition, once the food is placed in the mouth, normal digestive processes take over without being affected by the lack of gravity.

The next problem was the effect of weightlessness on the cardiovascular system, that is the heart and blood vessel system throughout the body. All types of reactions were possible in theory. In actual flight, a small and temporary amount of pooling of blood in the veins of the legs has occurred, but it is not serious nor does it appear to affect the performance of the pilot. For all pilots weightlessness has been a pleasant experience. All the senses such as sight and hearing perform normally during space flight. There has been no hallucination, no blackout or any other medical phenomena which might have an effect on Man in space. We even experimented with drifting flight and whether the astronaut would become disoriented when he could not distinguish up from down or have the horizon of Earth for a reference.

But each time the answer seemed to be that a man could adapt as long as his basic needs for breathing oxygen and pressure were supplied.

***

SCIENTIFIC EXPERIMENTS

Man's role as a scientific observer and experimenter in space was another unknown in the program. Much of it was based on the ability of man to exist in space. It had to first be determined that he would be able to function normally and then the scientific benefits of the program could be explored. Man as an observer has proven his capabilities from the first orbital flight. The brightness, coloring, and height of the airglow layer was [sic] established. It was something a camera could not record nor could an unmanned satellite perform this mission. Man in space has the ability to observe the unknown and to try to define it by experiment. The particles discovered at sunrise by John Glenn were determined by Scott Carpenter to be coming from the spacecraft, and this analysis was confirmed by Schirra and Cooper.

We can send unmanned instrumented vehicles into space which can learn much about the space environment and the make up of the planets. However, the use of Man to aid in making the scientific observations will be invaluable. The old problem of what and how to instrument for the unknown can benefit greatly from Man's capability to pick and choose the time and types of experiments to be performed. We have learned much from the Mercury program through this quality of choice and we will continue to learn if man continues to be an important part of the system.

If we have learned more about space itself, we have also learned about Man's capabilities in space. Many experiments have been conducted which have yielded valuable information for future programs. Aside from aeromedical experiments, Man has been able to distinguish color in space, to spot objects at varying distances from the spacecraft, to observe high intensity lights on the ground, and to track objects near him. These observations provide valuable information in determining the feasibility of the rendezvous and navigation in Gemini and Apollo.

***

CONCLUSION

The manned space flight program has changed quite a few concepts about space, added greatly to our knowledge of the universe around us, and demonstrated that Man has a proper role in exploring it. There are many unknowns that lie ahead, but we are reassured because we are confident in overcoming them by using Man's capabilities to the fullest.

When we started the manned space program five years ago, there was a great deal of doubt about Man's usefulness in space. We have now come to a point which is exactly one hundred eighty degrees around the circle from that opinion. We now depend on Man in the loop to back up the automatic systems rather than using automatic systems alone to insure that the mission is accomplished.

We do not want to ignore the automatic aspects of space flight altogether. There must be a careful blending of Man and machine in future spacecraft which provides the formula for further success. By experience, we have arrived at what we think is a proper mixture of that formula. Man is the deciding element; but we cannot ignore the usefulness of the automatic systems. As long as Man is able to alter the decision of the machine, we will have a spacecraft that can perform under any known condition, and that can probe into the unknown for new knowledge.

When President Kennedy on May 25, 1961, announced his decision to go to the Moon, NASA was planning to accomplish the lunar landing mission by building an extremely large rocket called Nova and using it to send a large spacecraft directly to the surface of the Moon. In the aftermath of Kennedy's announcement, NASA planners took a hard look at what developing Nova would require, and decided to use one or more smaller—though still large—launch vehicles to carry out the Moon mission. This meant that there would have to be some form of spacecraft rendezvous as part of the mission. Also, the lunar landing mission would require a spaceflight of seven or more days, and the existing Mercury spacecraft could not support an astronaut for more than three days. A new spacecraft capable of longer-duration missions would be required to test what impacts on astronauts might result from being in space for a week or more.

These and other considerations led the NASA leadership by the end of 1961 to decide that there was a pressing need for an interim human spaceflight project between Project Mercury and Project Apollo. That project was originally named "Mercury Mark II," the designation indicating that this was the second iteration of Project Mercury, since the new project was to be based on an upgraded Mercury spacecraft. That spacecraft would support a two-man crew, and the new project was quickly christened Project Gemini. The word "Gemini" was Latin for twins, and its choice reflected the two-person crew. The stated goal of the new effort was to provide "a versatile system which may be used for extending the time of flight in space and for development of rendezvous techniques, but may be adapted to the requirements of a multitude of other space missions at a later date." A sole-source contract to develop the Gemini spacecraft was awarded to McDonnell Aircraft. Another converted intercontinental ballistic missile, the Titan II, *was selected as the Gemini launch vehicle.*

The Gemini project development plan was approved by NASA associate administrator Robert Seamans on December 6, 1961. This original plan assumed the returning spacecraft would land on the ground rather than in the ocean, given the cost and complexity of water recovery. This approach was later abandoned as its feasibility came under question.

Manned Spacecraft Center (NASA), "Project Development Plan for Rendezvous Development Utilizing the Mark II Two Man Spacecraft," December 8, 1961

***

PART I—PROJECT SUMMARY

This project development plan presents a program of manned space flight during the 1963–1965 time period. The program provides a versatile system which may be used for extending the time of flight in space and for development of rendezvous techniques, but may be adapted to the requirements of a multitude of other space missions at a later date. A two man version of the Mercury spacecraft would be used in conjunction with a modified Titan II booster. The Atlas–Agena B combination would be used to place the Agena B into orbit as the target vehicle in the rendezvous experiments. This use of existing or modified versions of existing hardware minimizes the necessity for new hardware development.

The proposed plan is based on extensive usage of Mercury technology and components for the spacecraft. Therefore, it is proposed to negotiate a sole-source cost-plus-fixed-fee contract with McDonnell Aircraft Corporation for the Mark II Mercury spacecraft.

The launch vehicle procurement will involve a continuation of present arrangements with the Air Force and General Dynamics–Astronautics for the Atlas launch vehicles, and the establishment of similar arrangements with the Martin Company for the Modified Titan II launch vehicles, and with the Lockheed Aircraft Corporation for the Agena stages.

A Project Office will be established to plan, direct and supervise the program. The manpower requirements for this office are expected to reach 179 by the end of Fiscal Year 1962.

The estimated cost of the proposed program will total about 530 million dollars.

PART II—JUSTIFICATION

Upon completion of Project Mercury the next step in the overall plan of manned space exploration is to gain experience in long duration and rendezvous missions. It is believed that the program presented here would produce such information and that it would complement other programs now underway while not interfering with their prosecution.

***

PART IV—TECHNICAL PLAN

1.0 Introduction

Project Mercury is an initial step in a long range program of manned exploration of space. The initial objectives of Project

Mercury have already been accomplished; therefore, it now becomes appropriate to consider the steps that should be taken to
insure immediate continuation of manned space flights following
the successful conclusion of this project. Therefore, a follow-on
project, after Project Mercury, is proposed which will provide a
continuing source of development information. In the execution
of the proposed project, maximum use will be made of vehicle
and equipment development which has already been accomplished for other programs.

2.0 Mission Objectives

The present Mercury spacecraft cannot be readily adapted to
other than simple orbital missions of up to about one day duration, with a corresponding limitation on the objectives of the
mission. The proposed project will allow the accomplishment of
a much wider range of objectives.

2.1 **Long Duration Flights.** Experience will be gained in extending the duration of flights beyond the 18 orbit capability of the present Mercury spacecraft. It is recognized that
for the longer missions a multiman crew is essential so that
the work load may be shared, both in time and volume.
There are many areas which require investigations so that the
multiman crew may be provided with a suitable environment
during the prolonged missions. This project will contribute
to the development of the flight and ground operational techniques and equipment required for space flights of extended
periods. These flights will also determine the physiological
and psychological reactions and the performance capabilities
of the new crew while being subjected to extended periods in
a space environment.

2.2 **Rendezvous.** The rendezvous and docking maneuver in
space may be compared to serial refueling in that it makes
possible the resupply of a vehicle in space and thus extends
its mission capabilities. This maneuver makes it possible to
put a much larger "effective" payload in space with a given
booster. Since most space projects are "booster limited" at
present, the development of techniques for getting the most
out of available boosters should undoubtedly be treated as
of highest priority. As the frequency of manned orbital

flights increases, there will be instances when orbital rescue, personnel transfer, and spacecraft repair will be highly desirable. To accomplish these missions development of orbital rendezvous techniques is mandatory. Among the problem areas which are involved in effecting a successful rendezvous and docking maneuver are the following:

2.2.1 **Launch Window.** The second vehicle involved in the rendezvous must be launched very close to a prescribed time if the operation is to be economical in terms of waiting time and propulsion requirements. This requires a major simplification of the countdown procedure and high reliability of equipment.

2.2.2 **Navigation.** Means must be developed for maneuvers in space, using information supplied by the navigation system.

2.2.3 **Guidance and Control.** Guidance and control techniques must be developed for maneuvers in space, using information supplied by the navigation system.

2.2.4 **Docking.** Rendezvous is not effective until the docking maneuver is accomplished. The space environment makes this operation quite a bit different from the same type of operation within the earth's atmosphere and hence considerable work in developing suitable techniques is to be expected.

2.3 **Controlled Land Landings.** Experience has shown that the magnitude of the effort required to deploy adequate naval forces for the recovery of the Mercury spacecraft at sea is such that any means for avoiding, or at least minimizing, this effort would be highly desirable. The sea has proved to be a more inhospitable environment for recovery than was originally envisioned. If space flights are to be accomplished on anything like a routine basis, spacecraft must be designed to alight on land at specified locations. This requires that the landing dispersion be reduced to a very low figure, and a satisfactory method of touchdown developed.

2.3.1 **Dispersion Control.** To effect control of the landing area, it is fundamental that an impact prediction be made available to the pilot and a means provided for controlling the spacecraft so the desired impact point can be reached.

2.3.2 **Landing Impact.** The attenuation of the impact loads which might result from a land landing of the Mercury spacecraft has presented a very considerable problem. Although it is estimated that in many cases the landing accelerations would be within tolerable limits, the random nature of the landing process has made it impossible to consider a sufficient variety of conditions that could be encountered so as to have adequate assurance of success. In order to guarantee safety in landing, the impact must be made at a relatively low velocity and in a selected area.

2.4 **Training.** Although much can be accomplished by ground simulation training, there does not seem to be any real substitute for actual experience in space. Thus, a by-product of this project would be to provide a means of increasing the number of astronauts who have had actual experience in space. A two-manned spacecraft will be an excellent vehicle for this purpose.

2.5 **Project Philosophy.** In general, the philosophy used in the conception of this project is to make maximum use of available hardware, basically developed for other programs, modified to meet the needs of this project. In this way, requirements for hardware development and qualification are minimized and timely implementation of the project is assured.

***

This project will provide a versatile spacecraft/booster combination which will be capable of performing a variety of missions. It will be a fitting vehicle for conducting further experiments rather than be the object of experiments. For instance, the rendezvous techniques developed for the spacecraft might allow its use as a vehicle for resupply or inspection of orbiting laboratories or space stations, orbital rescue, personnel transfer, and spacecraft repair.

***

The first Gemini launch with a crew aboard—Mercury veteran Gus Grissom and new astronaut John Young—took place on March 23, 1965. By this time, the Soviet Union in October 1964 had flown a mission with a three-person crew; then, five days before Grissom and Young's Gemini 3 mission, cosmonaut Alexey Leonov stepped outside his spacecraft on the first extravehicular activity, i.e., a spacewalk. This achievement raised the question of whether NASA should react by adding a spacewalk to the Gemini 4 mission, which was scheduled for a June 3 launch. After an intensive review, the NASA leadership agreed to include a spacewalk during the Gemini 4 mission. This approval came on May 25, just a little more than a week before the launch. On June 3 astronaut Ed White conducted an EVA, becoming the first U.S. spacewalker; he spent twenty-one minutes outside the spacecraft and was reluctant to return because of the excitement of the event.

L. W. Vogel (Executive Officer), Memorandum for the Record, "Top Management Meeting on Gemini 4 Extra-Vehicular Activity," June 8, 1965

On May 24, Mr. Webb, Dr. Dryden and Dr. Seamans met with Dr. Mueller and Dr. Gilruth in connection with extra-vehicular activities on the Gemini 4 flight scheduled to take place on June 3.

Concern was expressed about changing the pattern of the flight. Making changes at the last minute always injected the possibility of something being overlooked and not properly considered. Also, if the Gemini 4 flight had to be cut short for any reason, opening the hatch would be blamed. Extra-vehicular activity in Gemini 4 was too obvious a reaction to the Soviet spectacular in this regard.

On the other hand, it was pointed out that suit development to permit extra-vehicular activity was part of the Gemini 4 program all along. Extra-vehicular activity had been originally planned for Gemini 4. One of the basic objectives of extra-vehicular activity was to be able to evaluate the possible utilization of man in space to carry out experiments, repair and adjust scientific satellites, and anything else that would require man to be outside of the spacecraft. The large antenna program was noted as one experiment which would require extra-vehicular activities by man.

It was then stated that there was no questioning of the propriety of having extra-vehicular activity in the Gemini program, but what was being questioned was it being performed on the second manned flight in the program. Since it was not essential to the basic mission of the Gemini 4 flight, which was to check out the reliability of the spacecraft and its systems for a 4-day period, our space posture might suffer if the 4-day period did not materialize.

The counter argument continued with comment about the great concern for the welfare of the astronauts and the fact that in the Gemini 3 flight we had a complete check on all systems. We have confidence in the spacecraft and the astronauts have trained for extra-vehicular activity and, if nothing than for morale purposes, they shouldn't do anything less than what they can do and have been trained to do. Extensive tests had been conducted under zero-gravity conditions in a K-135. The astronauts practiced getting in and out of the spacecraft under zero-gravity conditions a sufficient number of times so as to build up about an hour of experience. Also, it was pointed out that if we don't accomplish extra vehicular activity (EVA) in Gemini 4 then we must do it on Gemini 5. It is a logical extension of the Gemini program to do EVA on Gemini 4. If EVA is successful on GT-4, we will not do it on Gemini 5. If a decision were made today not to have EVA on GT-4, then we could do it on GT-5. However, it would be more of a compromise of the program to do EVA on GT-5 than on GT-4 because of the many other things programmed for GT-5.

The question was raised as to what risk we would be taking on a possible short Gemini 4 flight because of EVA and not finding out as much as we should find out about weightlessness as a problem. Weightlessness can be a problem, even in G-4, and we presumably will be concentrating on this problem in G-5. To this question it was noted that Dr. Berry said that there were no reservations about weightlessness being a problem over a 4-day period. There is no indication that 4 days of weightlessness will hurt man; therefore, this is not a great problem to be considered in the Gemini 4 flight. However, in connection with the Gemini 5 flight of 7 days there possibly are some reservations, primarily because no one has been in space for that period of time. Some medical experts feel that there will be a risk, others do not. Probably a problem just as pressing as the weightlessness problem is the problem of confinement for 7 days or longer periods.

The question was raised again as to the element of risk to complete the 4-day Gemini flight because of EVA. The reply was that the added risk was simply having to depressurize the spacecraft, open the hatch, seal the hatch, and repressurize the spacecraft. These procedures, involving various systems and sub-systems, of course add a degree of risk because of a possible failure. But these procedures have been done hundreds of times with no failure. Nevertheless, there is always a risk that something will not work, but this is a small risk.

It was noted that one cannot justify EVA in Gemini 4 just because the Russians did it, and one cannot justify EVA in Gemini 4 just because you want to get film out of the Agena rendezvous vehicle on a later Gemini flight. In rebuttal, it was commented that the main reason for EVA in the Gemini program is to further develop the role of man in space. The sophistication of equipment that we put into space is getting ahead of the sophistication of experiments we can do. Experiment sophistication can be increased through the use of man in space, but the use of man in space must be checked out by EVA. The determination as to whether man in space by extra-vehicular activity can repair things, can calibrate satellites, etc. should be looked upon as a significant step forward and not as a stunt.

***

To a comment that in the eyes of the public Gemini 4 would be a success with EVA, a statement was made that Gemini 4 with EVA might not necessarily be considered a success in the eyes of the decision makers. As a guide to risk taking, it was suggested that if there was a 90% chance to have a Gemini 4 flight for 4 days and that with EVA this chance would be only 89%, then we should risk 1% less chance for a 4-day flight for what can be gained from EVA. However, if a chance for a 4-day flight would be only 80% with EVA, then this additional 10% possibility for not having a 4-day flight would not be an adequate trade-off to be gained by EVA and we should not undertake it on Gemini 4.

It was noted that there was no comparison between the risk between the first Mercury flight and the Gemini 4 flight. It was recalled how the Air Force had admonished against the first Mercury flight, but NASA top management decided to go ahead because this flight was absolutely essential to the program. If we take into consideration the risks still inherent in using the rocket as a means of propulsion, then every time we use this means of propulsion we should find out everything that can be found out on the flight.

It was noted that we should not be too concerned about the public reaction in determining what is the best course of action. The decision as to whether or not there would be EVA on Gemini 4 should be made in the light of what is best for the program and should not be influenced by possible public reaction.

After the foregoing discussion, the concern was still raised that the importance of Gemini 4 was to check out the reliability of the spacecraft for 4 days and project this reliability for 7 days. EVA therefore might jeopardize getting everything we should get from Gemini 4. If Gemini 4 does not go for 4 days, then we are in a very difficult position for 7 days on Gemini 5 and presumably we could not go for 7 days on Gemini 5. The real question is whether or not EVA is important enough in view of the risk, no matter how slight, of jeopardizing a 4-day Gemini 4 flight and jeopardizing a 7-day Gemini 5 flight.

Then it was pointed out that if you look at the entire program, EVA is more logical for Gemini 4. If Gemini 4 lasts 3 days then we should not be concerned about spacecraft reliability for 7 days. The basic problems are really to check-out confinement and weightlessness. Therefore, Gemini 5 is more important than Gemini 4 and if there is any chance of reducing total flight time due to EVA, EVA then logically should be accomplished on Gemini 4 rather than on Gemini 5. Every guarantee was given to top management that if EVA were approved for Gemini 4, very firm and adequate instructions would be given covering the procedure.

Mr. Webb, Dr. Dryden and Dr. Seamans then gave careful consideration to the discussions they had with Dr. Mueller and Dr. Gilruth. In their opinion it was important, whatever the decision, that there be an adequate explanation to the public to avoid any unnecessary misunderstanding and to minimize any adverse reactions. There was a strong feeling to ratify EVA for Gemini 4 in order to get the maximum out of the flight. There was unanimity in that EVA eventually would be carried out, but there was some reservation as to whether or not it was the best judgment to have EVA on Gemini 4 as a risk beyond that which has to be taken. It was concluded that Dr. Seamans would discuss the matter further with Dr. Mueller and Dr. Gilruth, in view of the discussions which took place, and that if he did not care to press for EVA on Gemini 4, such EVA would not be undertaken. However, if the final discussion led Dr. Seamans to press for EVA in Gemini 4, then it would be unanimously approved for the flight.

NOTE: Following the meeting, a memorandum from Dr. Seamans to Mr. Webb, dated May 24, 1965, recommending EVA for the Gemini 4 flight was approved by Mr. Webb and Dr. Dryden.

There were eight more Gemini missions between August 1965 and November 1966. The Gemini 5 crew stayed in orbit for seven days; the Gemini 7 mission kept its crew aloft for almost fourteen days. After several failed attempts to rendezvous and dock the Gemini spacecraft with an Agena target vehicle, the Gemini 8 mission succeeded in such a docking. However, almost immediately after that docking, the Gemini/Agena combination started unexpectedly to rotate, forcing the crew—David Scott and Neil Armstrong—to undock. The rotation continued and speeded up to a dangerous level, with the possibility of the crew blacking out. Only quick action by the crew, particularly Armstrong, stopped the rotation, but those actions required an early termination of the mission. The first totally successful rendezvous and docking took place on the Gemini 10 mission.

Spacewalks took place on the Gemini 9, 10, and 11 missions, but there were problems with each. Only on the final mission, Gemini 12 in November 1966, was Edwin "Buzz" Aldrin able to spend more than five hours outside the spacecraft without problems.

The lessons learned from the Gemini program proved critical to the success of Apollo and the overall progress of human spaceflight. The program succeeded in accomplishing what had been intended for it from the outset, and then some. It demonstrated the capability of Americans to undertake long-duration

space missions. It provided the opportunity to develop rendez-
vous and docking techniques that served NASA's programs
well into the future. It pioneered the ability to leave the space-
craft and perform work outside in an extravehicular activity
(EVA).

This knowledge is summarized in the next two documents, ex-
plaining the results of the Gemini program. While the overview
given by NASA's Robert Seamans, by that point deputy admin-
istrator after the death of Hugh Dryden, is a straightforward
summary of the Gemini program, George Mueller, associate ad-
ministrator for manned spaceflight, eloquently lays out Gemini's
achievements in a broader context.

Robert C. Seamans Jr. (Deputy Administrator, NASA), Memorandum for Associate Administrators, Assistant Administrators, and Field Center Directors, NASA, "Gemini Program; Record of Accomplishments," January 17, 1967, with attached: "Project Gemini Summary"

The Gemini flight program, concluded on November 15, 1966, succeeded in accomplishing all of its pre-planned objectives, some of them several times over. As can be expected in any complex developmental-flight program, some of the individual flight missions experienced difficulties. The successful demonstration that these difficulties could be overcome in later missions is a tribute to the program organization, personnel directly involved, and to NASA . . .

PROJECT GEMINI SUMMARY

With the splashdown of Gemini 12 with astronauts Lovell and Aldrin aboard on November 15, 1966, the Gemini Project came to a successful conclusion. All Gemini Project objectives, including Extravehicular Activity and combined vehicle maneuvers, which were added after the project began, were fully accomplished many times over.

Rendezvous: Ten separate rendezvous were accomplished, using seven different techniques ranging from visual/manual control to ground/computer controlled rendezvous.

Docking: Nine dockings with four different Agenas were performed.

Docked Vehicle Maneuvers: Both Gemini X and Gemini XI demonstrated extensive maneuvers and a new altitude record was set on Gemini XI when the Agena target carried astronauts Conrad and Gordon 851 miles above the earth.

Extra-vehicular Activity: EVA was conducted on five separate Gemini Missions and during ten separate periods. Total EVA time during the Gemini Project was 12 hrs, 22 min. of which a record time of 5 hours and 37 minutes of EVA was performed by Aldrin on Gemini XII.

Long Duration Flight: Gemini VII demonstrated man's ability to stay in space continuously for up to 14 days; Gemini V for 8 days, and two other missions for 4 days.

Controlled Reentry: Landing accuracies of a few miles from the aim point were demonstrated on every Gemini manned mission except Gemini V.

Conduct Scientific and Technological Experiments: Every manned Gemini mission (Gemini III through XII) conducted many experiments. In total 43 experiments were conducted successfully.

Prior to each Gemini mission, individual primary mission objectives were selected which, if accomplished, would provide full advancement of the project. Accomplishment of these primary objectives were mandatory for stating the mission to be successful. To retain the flexibility to capitalize on success, secondary objectives were also assigned—as many as appeared feasible within the capability of the equipment and the time and experience of the astronauts.

Of the 14 Gemini mission attempts, 10 missions accomplished all of the primary mission objectives specified before the launch. The four unsuccessful missions and the reasons why they could not accomplish all of their primary objectives follows:

UNSUCCESSFUL MISSIONS	REASONS
GEMINI VI	The Agena Target Vehicle exploded. The Gemini 6 spacecraft was successfully rendezvoused with the Gemini 7 spacecraft later during the Gemini VI-A mission.
GEMINI VIII	An Orbit Maneuvering Thruster malfunction which ruled out a stated primary objective: EVA.
GEMINI IX	An Atlas booster failure drove the Agena into the Atlantic, and the Gemini 9 spacecraft was not launched until later during the Gemini IX-A mission.
GEMINI IX-A	The shroud did not come loose from the Augmented Target Docking Adapter, precluding docking—a specified primary objective for the mission.

George Mueller (Associate Administrator for Manned Space Flight), "Introduction," Gemini Summary Conference, NASA SP-138, February 1–2, 1967

The Gemini Program is over. . . . As is true in any undertaking of this magnitude, involving many diverse organizations and literally thousands of people, a vital element of the Gemini success may be traced to teamwork. In the purest definition of the word, wherein individual interests and opinions are subordinate to the unity and efficiency of the group, the Gemini team has truly excelled . . .

The Gemini Program was undertaken for the purpose of advancing the United States manned space-flight capabilities during the period between Mercury and Apollo. Simply stated, the Gemini objectives were to conduct the development and test program necessary to (1) demonstrate the feasibility of long duration space flight for at least that period required to complete a lunar landing mission; (2) perfect the techniques and procedures for achieving rendezvous and docking of two spacecraft in orbit; (3) achieve precisely controlled reentry and landing capability; (4) establish capability in the extravehicular activity; and (5) achieve the less obvious, but no less significant, flight and ground crew proficiency in manned space flight. The very successful flight program of the United States has provided vivid demonstration of the achievements in each of these objective areas.

The long-duration flight objective of Gemini was achieved with the successful completion of Gemini VII in December 1965. The progressive buildup of flight duration from 4 days with Gemini IV, to 8 days with Gemini V and 14 days with Gemini VII, has removed all doubts, and there were many, of the capability of the flight crews and spacecraft to function satisfactorily for a period equal to that needed to reach the lunar surface and return. Further, this aspect of Gemini provides high confidence in flight-crew ability to perform satisfactorily on much longer missions. The long-duration flights have also provided greater insight into, and appreciation of, the vital role played by the astronauts, the value of flexibility in mission planning and execution, and the excellent capability of the manned space-flight control system. As originally conceived, the Gemini Program called for completion of the long-duration flights with Gemini VII, which was accomplished on schedule.

One of the more dramatic achievements has been the successful development of a variety of techniques for the in-orbit rendezvous of two manned spacecraft. The preparation for this most complex facet of Gemini missions was more time consuming than any other. That it was performed with such perfection is a distinct tribute to the Gemini team that made it possible: the spacecraft and launch-vehicle developers and builders, the checkout and launch teams, the flight crews and their training support, and the mission-planning and mission-control people.

The ability to accomplish a rendezvous in space is fundamental to the success of Apollo, and rendezvous was a primary mission objective on each mission after Gemini VII. Ten rendezvous were completed and seven different rendezvous modes or techniques were employed. Nine different dockings of a spacecraft with a target vehicle were achieved. Eleven different astronauts gained rendezvous experience in this most important objective. Several of the rendezvous were designed to simulate some facet of an Apollo rendezvous requirement. The principal focus of the rendezvous activities was, however, designed to verify theoretical determinations over a wide spectrum. Gemini developed a broad base of knowledge and experience in orbital rendezvous and this base will pay generous dividends in years to come.

A related accomplishment of singular importance to future manned space-flight programs was the experience gained in performing docked maneuvers using the target vehicle propulsion system. This is a striking example of Gemini pioneering activities—the assembly and maneuvering of two orbiting space vehicles.

The first attempt at extravehicular activity during Gemini IV was believed successful, and although difficulties were encountered with extravehicular activity during Gemini IX-A, X, and XI, the objective was achieved with resounding success on Gemini XII. This in itself is indicative of the Gemini Program in that lessons learned during the flight program were vigorously applied to subsequent missions. The extravehicular activity on Gemini XII was, indeed, the result of all that had been learned on the earlier missions.

The first rendezvous and docking mission, although temporarily thwarted by the Gemini VI target-vehicle failure, was accomplished with great success during the Gemini VII/VI-A mission. This mission also demonstrated the operational proficiency achieved by the program. The term "operational proficiency" as applied to Gemini achievements means far more than just the acceleration of production rates and compressing of launch schedules. In addition and perhaps more importantly, operational proficiency means the ability to respond to the unexpected, to prepare and execute alternate and contingency plans, and to maintain flexibility while not slackening the drive toward the objective. Time and again Gemini responded to such a situation in a manner that can only be described as outstanding.

A few comments are in order on what the Gemini accomplishments mean in terms of value to other programs. There is almost no facet of Gemini that does not contribute in some way to the Apollo Program. Aside from the actual proof testing of such items as the manned space-flight control center, the manned space-flight communications net, the development and perfection of recovery techniques, the training of the astronauts, and many others which apply directly, the Gemini Program has provided a high level of confidence in the ability to accomplish the Apollo Program objectives before the end of this decade. The Apollo task is much easier now, due to the outstanding performance and accomplishments of the Gemini team.

Similarly, the Apollo Applications Program has been inspired in large part by the Gemini experiments program, which has sparked the imagination of the scientific community. In addition to the contributions to Apollo hardware development which provide the basis for the Apollo Applications Program, it has been discovered, or rather proved, that man in space can serve many extremely useful and important functions. These functions have been referred to as technological fallout, but it is perhaps more accurate to identify them as accomplishments—that is, accomplishments deliberately sought and achieved by the combined hard labor of many thousands of people . . .

The Manned Orbiting Laboratory Program has been undertaken by the Department of Defense for the purpose of applying manned space-flight technology to national defense and is making significant use of the Gemini accomplishments. This may be considered as a partial repayment for the marvelous support that NASA has received and continues to receive from the DOD. The success of the NASA programs is in no small measure due to the direct participation of the DOD in all phases of the manned space-flight program. This support has been, and will continue to be, invaluable.

The combined Government/industry/university team that makes up the manned space-flight program totals about 240,000 people. In addition, thousands more are employed in NASA unmanned space efforts, and in programs of the Department of Defense, the Department of Commerce, the Atomic Energy Commission, and other agencies involved in total national space endeavors. These people, in acquiring new scientific knowledge, developing new techniques, and working on new problems with goals ever enlarge by the magnitude of their task, form the living, growing capability of this Nation for space exploration.

For the last quarter century, this Nation has been experiencing a technological revolution. Cooperative efforts on the part of the Government, the universities, the scientific community, and industry have been the prime movers. This cooperation has provided tremendous capability for technological research and development which is available now and which will continue to grow to meet national requirements of the future. The influence of this technological progress and prowess is, and has been, a deciding factor in keeping the peace. Preeminence in this field is an important instrument in international relations and vitally influences this country's dealings with other nations involving peace and freedom in the world. Political realities which can neither be wished away nor ignored make the capability to explore space a matter of strategic importance as well as a challenge to the scientific and engineering ingenuity of man. This Nation can no more afford to falter in space than it can in any earthly pursuit on which the security and future of the Nation and the world depend.

The space effort is really a research and development competition, a competition for technological preeminence which demands and creates the quest for excellence.

The Mercury program, which laid the groundwork for Gemini and the rest of this Nation's manned space-flight activity, appears at this point relatively modest. However, Mercury accomplishments at the time were as significant to national objectives as the Gemini accomplishments are today and as those that are planned for Apollo in the years ahead.

That these programs have been, and will be, conducted in complete openness with an international, real-time audience makes them all the more effective. In this environment, the degree of perfection achieved is even more meaningful. Each person involved can take richly deserved pride in what has been accomplished. Using past experience as a foundation, the exploration of space must continue to advance. The American public will not permit otherwise, or better yet, history will not permit otherwise.

CHAPTER 3

"ONE SMALL STEP . . . ONE GIANT LEAP"[5]

Projects Mercury and Gemini would lead to NASA's crowning achievement in human spaceflight: Project Apollo, the remarkable U.S. space effort that sent twelve astronauts to the surface of the Earth's Moon. In any discussion of space exploration, the Moon had never been far from hand. The NASA Long Range Plan published in December 1959 had identified missions to land on the Moon as the long-range goal of the NASA human spaceflight program. Those landings were to take place "beyond 1970." Objectives for the 1965 to 1967 time period, after the completion of Project Mercury (Project Gemini had not yet been conceived), were the first launches "in a program leading to manned circumlunar flight and to [a] permanent near-earth space station."

By mid-1960, NASA's thinking about the intermediate steps in human spaceflight had matured to the point that the space agency called together representatives of the emerging space industry to share that thinking. At a "NASA-Industry Program Plans Conference" held in Washington on July 28 and 29, 1960, George Low, the head of human spaceflight at NASA's Washington headquarters, told the audience that "our present planning calls for the development and construction of an advanced manned spacecraft with sufficient flexibility to be capable of both circumlunar flight and useful earth-orbital missions. In the long range, this spacecraft should lead toward manned landings on the moon and planets, and toward a permanent manned space station. This advanced manned space flight program has been named 'Project Apollo.'" The name Apollo had been suggested by Low's boss, NASA's director for space flight programs Abe Silverstein, in early 1960.

NASA, and particularly George Low, continued to move forward in the second half of 1960 in planning what might follow Project Apollo, which at that point had only Earth orbit and circumlunar flight as its goals. On October 17, 1960, Low informed Silverstein that "it has become increasingly apparent that a preliminary program for manned lunar landings should be formulated. This is necessary in order to provide a proper justification for Apollo, and to place Apollo schedules and technical plans on a firmer foundation." To undertake this planning, Low formed a small working group of NASA Headquarters staff. The results of that group's work were soon to form the technical basis for a presidential commitment to send Americans to the Moon "before this decade is out."

That NASA was planning advanced human spaceflight missions, including ones to land people on the Moon, soon came to the attention of President Eisenhower and his advisers. The president asked his science adviser, George Kistiakowsky, to organize a study of "the goals, the missions and the costs" of the human spaceflight program that NASA had in mind. To carry out such a study, Kistiakowsky established an "Ad Hoc Committee on Man-in-Space" chaired by Brown University professor Donald Hornig. The Hornig Committee issued its report on December 16, 1960, and briefed it to President Eisenhower a few days later. Eisenhower has been quoted as saying at that time that "he couldn't care less whether a man ever reaches the Moon," and when a comparison was made to Queen Isabella's willingness to finance the voyages of Christopher Columbus, Eisenhower replied that "he was not about to hock his jewels" to send men to the Moon.

President's Science Advisory Committee, "Report of the Ad Hoc Panel on Man-in-Space," December 16, 1960

1. INTRODUCTION

We have been plunged into a race for the conquest of outer space. As a reason for this undertaking some look to the new and exciting scientific discoveries which are certain to be made. Others feel the challenge to transport man beyond frontiers he scarcely dared dream about until now. But at present the most impelling reason for our effort has been the international political situation which demands that we demonstrate our technological capabilities if we are to maintain our position of leadership. For all of these reasons we have embarked on a complex and costly adventure. It is the purpose of this report to clarify the goals, the missions and the costs of this effort in the foreseeable future, particularly with regard to the man-in-space program . . .

2. THE MAN-IN-SPACE PROGRAM

The initial American attempt to launch a manned capsule into orbital flight, Project Mercury, is already well advanced. It is a somewhat marginal effort, limited by the thrust of the Atlas booster. It has as its goal the launching of a one man capsule into orbit around the earth and its successful return to earth. The fact that the thrust of any available American booster is barely sufficient for the purpose means that it is difficult to achieve a high probability of a successful flight while also providing adequate safety for the Astronaut. Achieving reliability on both accounts will strain our capabilities. A difficult decision will soon be necessary as to when or whether a manned flight should be launched. The chief justification for pushing Project Mercury on the present time scale lies in the political desire either to be the first nation to send a man into orbit, or at least to be a close second.

**

None of the boosters now planned for development are capable of landing on the moon with sufficient auxiliary equipment to return the crew safely to earth. To achieve this goal, a new program much larger than Saturn will be needed. It is likely to take one of three forms:

1. An all-chemical liquid-fueled rocket, the Nova, might be developed to take the trip directly. It would require a booster with about 6 times the thrust of the Saturn and utilizing either kerosene or hydrogen-oxygen. The upper stage of the Nova would require hydrogen-oxygen and at least one stage would probably be an existing stage from the Saturn development program.

2. If a suitable nuclear upper stage could be developed, the Nova vehicle could conceivably become a combination chemical-nuclear system. This system would still require the development of a first stage chemical booster with thrust of the same order of magnitude as that described for the all-chemical system. If the nuclear development should be as successful as its proponents hope, it might open the way for future developments beyond the possibilities envisioned for chemical rockets. However, a sound decision on the promise of nuclear rockets cannot be made until about 1963.

3. Rendezvous techniques, utilizing either Saturn C-2 vehicles or some type of advanced Saturn vehicles, could be employed to lift into an earth orbit the hardware and fuel necessary to perform the manned lunar landing mission. In this system, a series of vehicles would be launched into a temporary earth orbit where they would rendezvous to enable fueling of the spacecraft and, if necessary, assembly of the component parts of the spacecraft. This spacecraft would then be used to transport the manned payload to the moon and thence back to Earth. These techniques will require considerable development, and are at present only in a preliminary study phase.

It is clear that any of the routes to land a man on the moon require a development much more ambitious than the present Saturn program. Not only must much bigger boosters probably be developed, but rockets and guidance mechanisms for the safe landing and then for return from moon to earth by means of additional rockets must be developed and tested. Nevertheless, it must be pointed out that this new, major step is implicit in undertaking the proposed manned Saturn program, for the first really big achievement of the man-in-space program would be the lunar landing. The succeeding step, manned flight to the vicinity of Venus or Mars, represents a problem an order of magnitude greater than that involved in the manned lunar landing. Not only does it appear to be insoluble in terms of chemical rockets, thus requiring the development of suitable nuclear rockets or nuclear-powered electric propulsion devices, but it also poses serious problems in terms of life support and radiation shielding for journeys requiring times ranging from many months to years.

***

4. RELATION BETWEEN MANNED AND UNMANNED SPACE EXPLORATION

Certainly among the major reasons for attending the manned exploration of space are emotional compulsions and national aspirations. These are not subjects which can be discussed on technical grounds. However, it can be asked whether the presence of a man adds to the variety or quality of the observations which can be made from unmanned vehicles, in short whether there is a scientific justification to include man in space vehicles.

It is said that an astronaut's judgment, decision-making capability and resourcefulness can increase the probability of successful accomplishment of a space mission and expand the variety and quality of observations performed. On the other hand, man's senses can be satisfactorily duplicated at remote locations by the use of available instrumentation and advances in the state of the art are continually increasing the ability to transmit information back to a central receiving point. With such an instrumented system, the decisions requiring man's mental capabilities can be performed by many men in a normal environment and with the aid of elaborate computational aids, where necessary.

The following considerations seem pertinent:

1.   Information from unmanned flights is a necessary prerequisite to manned flight.
2.   The degree of reliability that can be accepted in the entire mechanism is very much less for unmanned than for manned vehicles. As the systems become more complex this may make a decisive difference in what one dares to undertake at any given time.
3.   From a purely scientific point of view it should be noted that unmanned flights to a given objective can be undertaken much earlier. Hence repeated observations, changes of objectives and the learning by experience are more feasible.

It seems, therefore, to us at the present time that man-in-space cannot be justified on purely scientific grounds, although more thought may show that there are situations for which this is not true. On the other hand, it may be argued that much of the motivation and drive for the scientific exploration of space is derived from the dream of man's getting into space himself.

***

6. CONCLUSIONS

1. The first major goal of the man-in-space program is to orbit a man about the earth. It will cost about 350 million dollars.

2. The next goal, of an intermediate nature, is the manned circumnavigation of the moon. It will cost about 8 billion dollars.

3. The second *major* goal, landing on the moon, can only be achieved about 1975 after an additional national expenditure in the vicinity of 26 to 38 billion dollars.

4. The Saturn program is a necessary intermediate step toward manned lunar landing but must be followed by a much bigger development before manned lunar landing is possible.

5. The unmanned program is a necessary prerequisite to a manned program. Even if there were no manned program, the unmanned program might yield as much scientific knowledge and on this basis would be justified in its own right.

6. Even if there were no man-in-space program, Saturn C-2 is still a minimum vehicle for close-up instrumented study of Venus and Mars, for unmanned trips to more distant planets, and for putting roving vehicles on the surface of the moon.

7. Manned trips to the vicinity of Venus or Mars are not yet foreseeable.

When Dwight Eisenhower left office on January 20, 1961, the future of NASA's program of human spaceflight was extremely uncertain. There were no funds in Eisenhower's final budget proposal to support Project Apollo, and it was known that the incoming president, John F. Kennedy, was receiving advice skeptical of the value of launching humans into space.

As he entered the White House, Kennedy was aware that he would be faced with decisions that would shape the future of U.S. space efforts. The first order of business for the new president was to select someone to head NASA. His choice was veteran Washington operator James E. Webb. Once Webb arrived at NASA, a first task was to review the agency's proposed budget for FY1962 that had been prepared by the outgoing Eisenhower administration. In doing so, Webb and his associates came to the conclusion that NASA's planning had been too conservative, and that the milestones included in the agency's ten-year plan should be accelerated.

One source of input into this conclusion was the February 7 final report of George Low's Working Group, the first fully developed plan for how NASA might send men to the Moon. The report is notable for its estimate that a lunar landing program could be undertaken with the addition of $700 million per year to the NASA budget over a ten-year period. The report optimistically concluded that "no invention or breakthrough" would be needed to accomplish a lunar landing. The report recognized that NASA would need additional institutional capacity to carry out a Moon program. It also anticipated that recovery and reuse of launch vehicles would be key to lowering the costs of space exploration; that would not happen until 2017, fifty-six years in the future.

George M. Low, "A Plan for a Manned Lunar Landing," February 7, 1961

INTRODUCTION

In the past, man's scientific and technical knowledge was limited by the fact that all of his observations were made either from the earth's surface or from within the earth's atmosphere. Now man can send his measuring equipment on satellites beyond the earth's atmosphere and into space beyond the moon on lunar and planetary probes. These initial ventures into space have already greatly increased man's store of knowledge. In the future, man himself is destined to play a vital and direct role in the exploration of the moon and of the planets. In this regard, it is not easy to conceive that instruments can be devised that can effectively and reliably duplicate man's role as an explorer, a geologist, a surveyor, a photographer, a chemist, a biologist, a physicist, or any of a host of other specialists whose talents would be useful.

In all of these areas, man's judgment, his ability to observe and to reason, and his decision-making capabilities are required.

The initial step in our program for the manned exploration of space is Project Mercury. This Project is designed to put a manned satellite into an orbit more than 100 miles above the earth's surface, let it circle the earth three times, and bring it back safely. From Project Mercury we expect to learn much about how man will react to space flight, what his capabilities may be, and what should be provided in future manned spacecraft to allow man to function usefully. Such knowledge is vital before man can participate in other, more difficult, space missions.

Project Mercury is the beginning of a series of programs of ever-increasing scope and complexity . . .

The next step after Mercury is Project Apollo. The multi-manned Apollo spacecraft will provide for the development and exploitation of manned space flight technology in earth orbit; it also provide the initial step in a long-range program for the manned exploration of the moon and the planets.

In this paper we will focus on a major milestone in the program for manned exploration of space—lunar landing and exploration. This milestone might be subdivided into two phases:

1. Initial manned landing, with return to earth;
2. Manned exploration.

This report will be limited to a discussion of the initial manned lunar landing and return mission, with the clear recognition that it is a part of an integrated plan leading toward manned exploration of the moon.

An undertaking such as manned lunar landing requires a team effort on an exceedingly broad scale . . . The basic capability is provided through the parallel development of a spacecraft and a launch vehicle. Both of these developments must proceed in an orderly fashion, leading to hardware of increasing capability. Supporting these developments are many other scientific and technical programs and disciplines . . . The implementation of the manned spacecraft program requires information that will be obtained in the unmanned spacecraft and life science programs. The development of launch vehicle capability requires new engines, techniques to launch from earth orbit, and might include launch vehicle recovery developments. Both the spacecraft and the launch vehicle programs can progress only as new knowledge is obtained through advanced research . . .

NASA RESEARCH

Already there exists a large fund of basic scientific knowledge, as a result of the advanced research of the past several years, which permits confidence that the technology required for manned lunar flight can be successfully developed. It would be misleading to imply that all of the major problems are now clearly foreseen; however, there is an acute awareness of the magnitude of the problems. The present state of knowledge is such that no invention or breakthrough is believed to be required to insure the overall feasibility of safe manned lunar flight.

***

It is proposed, therefore, that a vehicle larger than the Saturn C-2 be phased into the launch vehicle program in an orderly fashion following the Saturn development. Such a launch vehicle, called Nova, would use a cluster of 1,500,000 pound thrust F-1 engines in its booster stage. The exact number of F-1 engines will have to be determined later, when a more complete definition of Nova missions is in hand. Nova might be sufficiently large to permit a manned lunar landing with a single launching directly from earth. Or, although substantially larger than the Saturn C-2, it might still not be large enough to approach the moon directly from earth; in this case it would materially reduce the number of rendezvous operations needed in earth orbit for each lunar mission.

***

Use of the Nova-class vehicle offers the possibility of greatly reducing the required number of launchings from earth. It might be possible to provide mission capability without rendezvous with a four-engine Nova; with an eight-engine Nova, this type of mission capability is virtually assured . . .

It is possible that other propulsion developments could contribute to manned lunar flight capability. Examples are the use of large solid propellant rockets, or nuclear propulsion. In defining a

Nova configuration, consideration will be given to both of these types of propulsion. At the present time it appears that nuclear propulsion will not be sufficiently developed for the initial manned lunar landing; however, nuclear propulsion might be very desirable and economically attractive for later exploration of the moon.

***

SPACECRAFT DEVELOPMENT

The spacecraft development for the manned lunar landing mission will be an extension of the Apollo program. Before a spacecraft capable of manned circumlunar flight and lunar landing can be designed, a number of unknowns must be answered.

The two most serious questions are:

1. What are the effects on man of prolonged exposure to weightlessness?
2. How may man best be protected from radiation in space?

The entire spacecraft design, its shape and its weight, will depend to a great extent on whether or not man can tolerate prolonged periods of weightlessness. And, if it is determined that he cannot, then the required amount of artificial gravity, or perhaps of other forms of sensory stimulation, will have to be specified.

The spacecraft design and weight will also be greatly affected by the amount of radiation shielding required to protect a man. In this area, a clear definition of the pertinent types of radiation, and their effects on living beings, is needed.

***

Manned landings on the moon . . . could be made in the 1968–1971 time period. If orbital operations using the Saturn C-2 vehicles prove to be practicable for this mission, then it might be accomplished toward the beginning of this range of time. On the other hand, if the spacecraft becomes much more complex than now envisioned, and consequently much heavier, a Nova vehicle will most likely be required before man can be landed on the moon. In the latter event, the program goals may not be accomplished as quickly.

***

The average cost per year, over a ten year period, for the total program is of the order of $700,000,000.

***

An examination of the required NASA staffing to carry out this plan was not made as a part of this study. However, it must be recognized that neither Marshall Space Flight Center nor Space Task Group, as presently staffed, could fully support these programs. If the program is to be adopted, immediate consideration must be given to this problem.

CONCLUDING REMARKS

In preparing this plan for a manned lunar landing capability, it was recognized that many foreseeable problems will require solutions before the plan can be fully implemented. Yet, an examination of ongoing NASA programs in the areas of advanced research, life sciences, spacecraft development, and engine and launch vehicle development, has shown that solutions to all of these problems should be available in the required period of time.

Throughout the plan, allowances were made for foreseeable problems; but it must be recognized that unforeseeable problems might delay the accomplishment of this mission. Nevertheless, the plan is believed to be sound in that it requires, at each point in time, a minimum commitment of funds and resources until the needed background information is in hand. Thus, the plan does not represent a "crash" program, but rather it represents a vigorous development of technology . . .

In his Inaugural Address, delivered on a wintry January afternoon, President John F. Kennedy suggested to the leaders of the Soviet Union, "together let us explore the stars." In his initial thinking, Kennedy favored using space activities as a way of increasing the peaceful interactions between the United States and its Cold War adversary. Soon after he came to the White House, Kennedy directed his science adviser to undertake an intensive review to identify areas of potential U.S.-Soviet space cooperation, and that review continued for the first three months of the

*Kennedy administration, only to be overtaken by the need to re-
spond to the Soviet launch of Yuri Gagarin on April 12, 1961.
Soviet-U.S. cooperation in space was a theme that Kennedy was
to return to in subsequent years.*

*Kennedy had not yet made up his own mind about the future of
human spaceflight, and so in March he was unwilling to approve
NASA's request to restore funds for the Apollo spacecraft that
President Eisenhower had deleted from the NASA budget request;
the sense was that decisions on this issue would come during the
preparation of the FY1963 NASA budget at the end of 1961.*

*Events forced the president's hand much earlier than he had
anticipated. In the early-morning hours of April 12, word reached
the White House that the Soviet Union had successfully orbited
its first cosmonaut, Yuri Gagarin, and that he had safely returned to Earth. The USSR was quick to capitalize on the propaganda impact of the Gagarin flight; Nikita Khrushchev boasted,
"Let the capitalist countries catch up with our country!" In the
United States, both the public and the Congress demanded a response to the Soviet achievement.*

*President Kennedy called a meeting of his advisers for the late
afternoon of April 14 to discuss what that response might be.
NASA deputy administrator Hugh Dryden told the president
that to catch up with the Russians might require a crash program on the order of the Manhattan Project that developed the
atomic bomb; such an effort could cost as much as $40 billion.
After hearing the discussions of what might be done, Kennedy's
response was "when we know more, I can decide if it's worth it
or not. If someone can just tell me how to catch up. . . . There's
nothing more important." Three days later, the Bay of Pigs fi-
asco took place, and Kennedy had added incentive to avoid
looking weak on the global stage.*

*Kennedy had decided in December 1960 to make his vice pres-
ident, Lyndon Johnson, the chairman of the existing National
Aeronautics and Space Council. That council had been set up as
part of the 1958 Space Act, with the president as chair. Thus legislative action was needed to give the chairmanship to the vice
president. The president signed the legislation making this change
on April 20, and on that same day wrote a historic memorandum
to the vice president, asking him "as Chairman of the Space*

Council to be in charge of making an overall survey of where we stand in space." This memorandum, written eight days after the Gagarin flight, set out very clear requirements for the U.S. response, asking for a "space program which promises dramatic results in which we could win." This was a clear signal that Kennedy intended to enter, and win, a space race with the Soviet Union.

John F. Kennedy, Memorandum for Vice President, April 20, 1961

THE WHITE HOUSE

WASHINGTON

April 20, 1961

MEMORANDUM FOR

VICE PRESIDENT

In accordance with our conversation I would like for you as Chairman of the Space Council to be in charge of making an overall survey of where we stand in space.

1. Do we have a chance of beating the Soviets by putting a laboratory in space, or by a trip around the moon, or by a rocket to land on the moon, or by a rocket to go to the moon and back with a man. Is there any other space program which promises dramatic results in which we could win?

2. How much additional would it cost?

3. Are we working 24 hours a day on existing programs. If not, why not? If not, will you make recommendations to me as to how work can be speeded up.

4. In building large boosters should we put out emphasis on nuclear, chemical or liquid fuel, or a combination of these three?

5. Are we making maximum effort? Are we achieving necessary results?

I have asked Jim Webb, Dr. Weisner, Secretary McNamara and other responsible officials to cooperate with you fully. I would appreciate a report on this at the earliest possible moment.

As he carried out his review Johnson consulted not only government agencies but also individuals whom he respected. One of those consulted was Wernher von Braun. Although von Braun as director of NASA's Marshall Space Flight Center was formally several layers down in the NASA hierarchy, he had become a publicly acclaimed space advocate, and the vice president wanted his opinions unfiltered through von Braun's NASA bosses. Von Braun met with Johnson and his associates on April 27, and followed up that meeting with an April 29 letter. His statement that the United States had an "excellent chance" of being first in sending humans to the surface of the Moon and back safely to Earth played an influential role in selecting a lunar landing as the means of winning the space race.

Letter from Wernher von Braun to the Vice President of the United States, April 29, 1961

Dear Mr. Vice President:

This is an attempt to answer some of the questions about our national space program raised by the President in his memorandum to you dated April 20, 1961. I should like to emphasize that the following comments are strictly my own and do not necessarily reflect the official position of the National Aeronautics and Space Administration in which I have the honor to serve.

Question 1. Do we have a chance of beating the Soviets by putting a laboratory in space, or by a trip around the moon, or by a rocket to land on the moon, or by a rocket to go to the moon and back with a man? Is there any other space program which promises dramatic results in which we could win?

Answer: With their recent Venus shot, the Soviets demonstrated that they have a rocket at their disposal which can place 14,000 pounds of payload in orbit. When one considers that our own one-man Mercury space capsule weighs only 3900 pounds, it becomes readily apparent that the Soviet carrier rocket should be capable of

- launching *several* astronauts into orbit simultaneously. (Such an enlarged multi-man capsule could be considered and could serve as a small "laboratory in space.")

- soft-landing a substantial payload on the moon. My estimate of the maximum soft-landed net payload weight the Soviet rocket is capable of is about 1400 pounds (one-tenth of its low orbit payload). This weight capability is *not* sufficient to include a rocket for the *return flight* to earth of a man landed on the moon. But it is entirely adequate for a powerful radio transmitter which would relay lunar data back to earth and which would be *abandoned* on the lunar surface after completion of this mission. A similar mission is planned for our "Ranger" project, which uses an Atlas-Agena B boost rocket. The "semi-hard" landed portion of the Ranger package weighs 293 pounds. Launching is scheduled for January 1962.

The existing Soviet rocket could furthermore hurl a 4000 to 5000 pound capsule *around* the moon with ensuing re-entry into the earth atmosphere. This weight allowance must be considered marginal for a one-man round-the-moon voyage. Specifically, it would not suffice to provide the capsule and its occupant with a "safe abort and return" capability, a feature which under NASA ground rules for pilot safety is considered mandatory for all manned space flight missions. One should not overlook the possibility, however, that the Soviets may substantially facilitate their task by simply waiving this requirement.

A rocket about ten times as powerful as the Soviet Venus launch rocket is required to land a man on the moon and bring him back to earth. Development of such a super rocket can be circumvented by orbital rendezvous and refueling of smaller rockets, but the development of this technique by the Soviets would not be hidden from our eyes and would undoubtedly require several years (possibly as long or even longer than the development of a large direct flight super rocket).

Summing up, it is my belief that

a.   we do *not* have a good chance of beating the Soviets to a manned "*laboratory in space.*" The Russians could place it in orbit this year while we could establish a (somewhat heavier) laboratory only after the availability of a reliable Saturn C-1 which is in 1964.

b. we *have* a sporting chance of beating the Soviets to a soft-landing of a radio *transmitter station on the moon.* It is hard to say whether this objective is on their program, but as far as the launch rocket is concerned, they could do it at any time. We plan to do it with the Atlas-Agena B-Ranger #3 in early 1962.

c. we have a sporting chance of sending a 3-man crew *around the moon* ahead of the Soviets (1965/66). However, the Soviets could conduct a round-the-moon voyage earlier if they are ready to waive certain emergency safety features and limit the voyage to one man. My estimate is that they could perform this simplified task in 1962 or 1963.

d. we have an excellent chance of beating the Soviets to the *first landing of a crew on the moon* (including return capability, of course). The reason is that a performance jump by a factor 10 over their present rockets is necessary to accomplish this feat. While today we do not have such a rocket, it is unlikely that the Soviets have it. Therefore, we would not have to enter the race toward this obvious next goal in space exploration against hopeless odds favoring the Soviets. With an all-out crash program I think we could accomplish this objective in 1967/68.

***

Question 5. Are we making maximum effort? Are we achieving necessary results?

Answer: No, I do *not* think we are making maximum effort. In my opinion, the most effective steps to improve our national stature in the space field, and to speed things up would be to

- identify a few (the fewer the better) goals in our space program as objectives of highest national priority. (For example: Let's land a man on the moon in 1967 or 1968.)
- identify those elements of our present space program that would qualify as immediate contributions to this objective. (For example, soft landings of suitable

instrumentation on the moon to determine the environmental conditions man will find there.)

- put all other elements of our national space program on the "back burner."
- add another more powerful liquid fuel booster to our national launch vehicle program. The design parameters of this booster should allow a certain flexibility for desired program reorientation as more experience is gathered . . .

Summing up, I should like to say that in the space race we are competing with a determined opponent whose peacetime economy is on a wartime footing. Most of our procedures are designed for orderly, peacetime conditions. I do not believe that we can win this race unless we take at least some measures which thus far have been considered acceptable only in times of a national emergency.

On April 28, Johnson delivered an interim report to Kennedy on the results of his review of the space program. The report identified a lunar landing by 1966 or 1967 as the first dramatic space project in which the United States could beat the Soviet Union. The vice president argued that "leadership" should be the appropriate goal of U.S. efforts in space. That goal has remained influential in the years since.

Lyndon B. Johnson, Memorandum for the President, "Evaluation of Space Program," April 28, 1961

Reference is to your April 20 memorandum asking certain questions regarding this country's space program.

A detailed survey has not been completed in this time period. The examination will continue. However, what we have obtained so far from knowledgeable and responsible persons makes this summary reply possible.

Among those who have participated in our deliberations have been the Secretary and Deputy Secretary of Defense; General Schriever (AF); Admiral Hayward (Navy); Dr. von Braun (NASA); the Administrator, Deputy Administrator, and other top officials of NASA; the Special Assistant to the President on Science and

Technology; representatives of the Director of the Bureau of the Budget; and three outstanding non-Government citizens of the general public: Mr. George Brown (Brown & Root, Houston, Texas); Mr. Donald Cook (American Electric Power Service, New York, N.Y.); and Mr. Frank Stanton (Columbia Broadcasting System, New York, N.Y.).

The following general conclusions can be reported:

a. Largely due to their concentrated efforts and their earlier emphasis upon the development of large rocket engines, the Soviets are ahead of the United States in world prestige attained through impressive technological accomplishments in space.

b. The U.S. has greater resources than the USSR for attaining space leadership but has failed to make the necessary hard decisions and to marshal those resources to achieve such leadership.

c. This country should be realistic and recognize that other nations, regardless of their appreciation of our idealistic values, will tend to align themselves with the country which they believe will be the world leader—the winner in the long run. Dramatic accomplishments in space are being increasingly identified as a major indicator of world leadership.

d. The U.S. can, if it will, firm up its objectives and employ its resources with a reasonable chance of attaining world leadership in space during this decade. This will be difficult but can be made probable, even recognizing the head start of the Soviets and the likelihood that they will continue to move forward with impressive successes. In certain areas, such as communications, navigation, weather, and mapping, the U.S. can and should exploit its existing advance position.

e. If we do not make the strong effort now, the time will soon be reached when the margin of control over space and over men's minds through space accomplishments will have swung so far on the Russian side that we will not be able to catch up, let alone assume leadership.

f. Even in those areas in which the Soviets already have the capability to be first and are likely to improve upon such capability, the United States should make aggressive efforts as the technological gains as well as the international rewards are essential steps in eventually gaining leadership. The danger of long lags or outright omissions by this country is substantial in view of the possibility of great technological breakthroughs obtained from space exploration.

g.   Manned exploration of the moon, for example, is not only an achievement with great propaganda value, but it is essential as an objective whether or not we are first in its accomplishment—and we may be able to be first. We cannot leapfrog such accomplishments, as they are essential sources of knowledge and experience for even greater successes in space. We cannot expect the Russians to transfer the benefits of their experiences or the advantages of their capabilities to us. We must do these things ourselves.

h.   The American public should be given the facts as to how we stand in the space race, told of our determination to lead in that race, and advised of the importance of such leadership to our future.

i.   More resources and more effort need to be put into our space program as soon as possible. We should move forward with a bold program, while at the same time taking every practical precaution for the safety of the persons actively participating in space flights.

***

Johnson's review took place as NASA was preparing to make Alan Shepard the first American to enter space. During the same week, Kennedy asked Johnson to travel to Southeast Asia to get a sense of the situation there and determine whether direct U.S. military intervention was required. Decisions on what to recommend in response to Kennedy's April 20 memo thus needed to move quickly. Johnson wanted to get his final recommendations to the president before he left Washington on Monday, May 8; this meant that those preparing the basis for those recommendations would have to work over the weekend. The recommendations came in the form of a memorandum signed by James Webb and Secretary of Defense Robert McNamara, titled "Recommendations for Our National Space Program: Changes, Policies, Goals."

The Webb-McNamara memorandum called for an across-the-board acceleration of the U.S. space effort aimed at seeking leadership in all areas, not only in dramatic space achievements. As its centerpiece, the report recommended getting a man to the Moon before the end of the decade. A very expensive undertaking such as sending humans to the Moon was justified, according to Webb and McNamara, on grounds of national prestige in the context of the Cold War. Johnson later that day delivered the report to the president, without modification and with his concurrence. It is, in effect, the Magna Carta of Project Apollo.

Memorandum to the Vice President from James E. Webb (NASA Administrator), and Robert S. McNamara (Secretary of Defense), May 8, 1961, with attached: "Recommendations for Our National Space Program: Changes, Policies, Goals"

***

II. NATIONAL SPACE POLICY

The recommendations made in the preceding Section imply the existence of national space goals and objectives toward which these and other projects are aimed. Major goals are summarized in Section III. Such goals must be formulated in the context of a national policy with respect to undertakings in space. It is the purpose of this Section to highlight our thinking concerning the direction that such national policy needs to take and to present a backdrop against which more specific goals, objectives and detailed policies should, in our opinion, be formulated.

a. Categories of Space Projects

Projects in space may be undertaken for any one of four principal reasons. They may be aimed at gaining scientific knowledge. Some, in the future, will be of commercial or chiefly civilian value. Several current programs are of potential military value for functions such as reconnaissance and early warning. Finally, some space projects may be undertaken chiefly for reasons of national prestige.

The U.S. is not behind in the first three categories. Scientifically and militarily we are ahead. We consider our potential in the commercial/civilian area to be superior. The Soviets lead in space spectaculars which bestow great prestige. They lead in launch vehicles needed for such missions. These bestow a lead in capabilities which may some day become important from a military point of view. For these reasons it is important that we take steps to insure that the current and future disparity between U.S. and Soviet launch capabilities be removed in an orderly but timely way. Many other factors however, are of equal importance.

b. Space Projects for Prestige

All large scale space projects require the mobilization of resources on a national scale. They require the development and successful application of the most advanced technologies. They call for skillful management, centralized control and unflagging pursuit of long range goals. Dramatic achievements in space, therefore, symbolize the technological power and organizing capacity of a nation.

It is for reasons such as these that major achievements in space contribute to national prestige. Major successes, such as orbiting a man as the Soviets have just done, lend national prestige even though the scientific, commercial or military value of the undertaking may by ordinary standards be marginal or economically unjustified.

This nation needs to make a positive decision to pursue space projects aimed at enhancing national prestige. Our attainments are a major element in the international competition between the Soviet system and our own. The non-military, non-commercial, non-scientific but "civilian" projects such as lunar and planetary exploration are, in this sense, part of the battle along the fluid front of the cold war. Such undertakings may affect our military strength only indirectly if at all, but they have an increasing effect upon our national posture.

***

III. MAJOR NATIONAL SPACE GOALS

***

a. Manned Lunar Exploration

We recommend that our National Space Plan include the objective of manned lunar exploration before the end of this decade. It is our belief that manned exploration to the vicinity of and on the surface of the moon represents a major area in which international competition for achievement in space will be conducted. The orbiting of machines is not the same as the orbiting or landing of man. It is man, not merely machines, in space that captures the imagination of the world.

The establishment of this major objective has many implications. It will cost a great deal of money. It will require large efforts for a long time. It requires parallel and supporting undertakings which are also costly and complex. Thus, for example, the RANGER and SURVEYOR Projects and the technology associated with them must be undertaken and must succeed to provide the data, the techniques and the experience without which manned lunar exploration cannot be undertaken.

The Soviets have announced lunar landing as a major objective of their program. They may have begun to plan for such an effort years ago. They may have undertaken important first steps which we have not begun.

It may be argued, therefore, that we undertake such an objective with several strikes against us. We cannot avoid announcing not only our general goals but many of our specific plans, and our successes and our failures along the way. Our cards are and will be face up—theirs are face down.

Despite these considerations we recommend proceeding toward this objective. We are uncertain of Soviet intentions, plans or status. Their plans, whatever they may be, are not more certain of success than ours. Just as we accelerated our ICBM program we have accelerated and are passing the Soviets in important areas in space technology. If we set our sights on this difficult objective we may surpass them here as well. Accepting the goal gives us a chance. Finally, even if the Soviets get there first, as they may, and as some think they will, it is better for us to get there second than not at all. In any event we will have mastered the technology. If we fail to accept this challenge it may be interpreted as a lack of national vigor and capacity to respond.

**

Given Johnson's interest in the accelerated space program recommended in the May 8 memorandum, NASA Administrator Webb prepared a status review for him upon his return from

Southeast Asia. This document is an excellent example of Webb's expansive view of the possible impacts of Apollo. Webb was contemplating the mobilization of human and financial resources that would be required to accomplish the lunar landing program; he saw the program as an opportunity for building new centers of regional technical competence. The memo is also an example of Webb's ability to massage the egos of leading politicians, in this case the vice president.

James E. Webb (NASA Administrator), Memorandum for the Vice President, May 23, 1961

***

4. In preparing for the hearings on the original Kennedy submission before the House Appropriations Committee, and in other discussions with Congressman Thomas [Representative Albert Thomas was a Houston-area congressman who chaired the subcommittee controlling NASA's budget appropriation], Thomas made it very clear that he and George Brown [head of the Houston construction firm Brown & Root] were extremely interested in having Rice University make a real contribution to the effort, particularly in view of the fact that some research funds were now being spent at Rice, that the resources of Rice had increased substantially, and that some 300 acres of land had been set aside for Rice for an important research installation. On investigation, I find that we are going to have to establish some place where we can do the technology related to the Apollo program, and this should be on the water where the vehicle can ultimately be barged to the launching site. Therefore we have looked carefully at the situation at Rice, and at the possible locations near the Houston Ship Canal or other accessible waterways in that general area. George Brown has been extremely helpful in doing this. No commitments whatever have been made, but I believe it is going to be of great importance to develop the intellectual and other resources of the Southwest in connection with the new programs which the Government is undertaking . . . If

it were possible to get a combination where the out-in-front theoretical research were done by Berkner [Webb's good friend Lloyd Berkner, chair of the Space Science Board of the National Academy of Science] and his group around Dallas in such a way as to strengthen all the universities in the area, and if at the same time a strong engineering and technological center could be established near the water near Houston and perhaps in conjunction with Rice University, these two strong centers would provide a great impetus to the intellectual and industrial base of this whole region and would permit us to think of the country as having a complex in California running from San Francisco down through the new University of California installation at San Diego, another center around Chicago with the University of Chicago as a pivot, a strong Northeastern arrangement with Harvard, M.I.T., and like institutions participating, some work in the Southeast perhaps revolving around the research triangle in North Carolina . . . and with the Southwestern complex rounding out the situation. I am sure you know that the decisions relating to this must await the completion of the work on our program by the Congress, but I am convinced, and believe you should consider very carefully, that it will attract the kind of strong support that will permit the President and you to move the program on through the Congress with minimum political in-fighting. I think this is important in the present situation and particularly to avoid the kind of end-runs that some of our friends related to the Pentagon, directly or industrially, have pursued in the past.

***

On May 25, 1961, President John F. Kennedy announced his decision to go to the Moon in an address to a joint session of Congress and a national television audience. His speech, titled "Urgent National Needs," reflected the slow start of the administration in convincing Congress and the American public that major steps had to be taken to increase national security. After listing a number of national security–related initiatives he believed were required, at the end of his speech Kennedy turned to space, and called upon the country to commit to "achieving the goal, before this decade is out, of landing a man on the moon and returning him safely to the earth." Toward the end of this section of the speech, Kennedy deviated from his prepared text to emphasize the "heavy burden" that accepting such a commitment would entail; most accounts of the lunar landing decision do not include those portions of his speech.

John F. Kennedy, "Urgent National Needs," Address to a Joint Session of the Congress, May 25, 1961

* * *

IX. SPACE

Finally, if we are to win the battle that is now going on around the world between freedom and tyranny, the dramatic achievements in space which occurred in recent weeks should have made clear to us all, as did the Sputnik in 1957, the impact of this adventure on the minds of men everywhere, who are attempting to make a determination of which road they should take. Since early in my term, our efforts in space have been under review. With the advice of the Vice President, who is Chairman of the National Space Council, we have examined where we are strong and where we are not, where we may succeed and where we may not. Now it is time to take longer strides—time for a great new American enterprise—time for this nation to take a clearly leading role in space achievement, which in many ways may hold the key to our future on earth.

I believe we possess all the resources and talents necessary. But the facts of the matter are that we have never made the national decisions or marshaled the national resources required for such leadership. We have never specified long-range goals on an urgent time schedule, or managed our resources and our time so as to insure their fulfillment.

Recognizing the head start obtained by the Soviets with their large rocket engines, which gives them many months of lead-time, and recognizing the likelihood that they will exploit this lead for some time to come in still more impressive successes, we nevertheless are required to make new efforts on our own. For while we cannot guarantee that we shall one day be first, we can guarantee that any failure to make this effort will make us last. We take an additional risk by making it in full view of the world, but as shown by the feat of astronaut Shepard, this very risk enhances our stature when we are successful. But this is not merely a race. Space is open to us now; and our eagerness to share its meaning is not governed by the efforts of others. We go into space because whatever mankind must undertake, free men must fully share.

I therefore ask the Congress, above and beyond the increases I have earlier requested for space activities, to provide the funds which are needed to meet the following national goals:

First, I believe that this nation should commit itself to achieving the goal, before this decade is out, of landing a man on the moon and returning him safely to the earth. No single space project in this period will be more impressive to mankind, or more important for the long-range exploration of space; and none will be so difficult or expensive to accomplish. We propose to accelerate the development of the appropriate lunar space craft. We propose to develop alternate liquid and solid fuel boosters, much larger than any now being developed, until certain which is superior. We propose additional funds for other engine development and for unmanned explorations—explorations which are particularly important for one purpose which this nation will never overlook: the survival of the man who first makes this daring flight. But in a very real sense, it will not be one man going to the moon—if we make this judgment affirmatively, it will be an entire nation. For all of us must work to put him there.

Secondly, an additional 23 million dollars, together with 7 million dollars already available, will accelerate development of the Rover nuclear rocket. This gives promise of some day providing a means for even more exciting and ambitious exploration of space, perhaps beyond the moon, perhaps to the very end of the solar system itself.

Third, an additional 50 million dollars will make the most of our present leadership, by accelerating the use of space satellites for world-wide communications.

Fourth, an additional 75 million dollars—of which 53 million dollars is for the Weather Bureau—will help give us at the earliest possible time a satellite system for world-wide weather observation.

Let it be clear—and this is a judgment which the Members of the Congress must finally make—let it be clear that I am asking the Congress and the country to accept a firm commitment to a new course of action—a course which will last for many years and carry very heavy costs: 531 million dollars in fiscal '62—an estimated seven to nine billion dollars additional over the next five years. If we are to go only half way, or reduce our sights in the face of difficulty, in my judgment it would be better not to go at all.

Now this is a choice which this country must make, and I am confident that under the leadership of the Space Committees of the Congress, and the Appropriating Committees, that you will consider the matter carefully.

It is a most important decision that we make as a nation. But all of you have lived through the last four years and have seen the significance of space and the adventures in space, and no one can predict with certainty what the ultimate meaning will be of mastery of space.

I believe we should go to the moon. But I think every citizen of this country as well as the Members of the Congress should consider the matter carefully in making their judgment, to which we have given attention over many weeks and months, because it is a heavy burden, and there is no sense in agreeing or desiring that the United States take an affirmative position in outer space, unless we are prepared to do the work and bear the burdens to make it successful. If we are not, we should decide today and this year.

This decision demands a major national commitment of scientific and technical manpower, materiel and facilities, and the possibility of their diversion from other important activities where they are already thinly spread. It means a degree of dedication, organization and discipline which have not always characterized our research and development efforts. It means we cannot afford undue work stoppages, inflated costs of material or talent, wasteful interagency rivalries, or a high turnover of key personnel.

New objectives and new money cannot solve these problems. They could in fact, aggravate them further—unless every scientist, every engineer, every serviceman, every technician, contractor, and civil servant gives his personal pledge that this nation will move forward, with the full speed of freedom, in the exciting adventure of space.

Congress quickly and without significant opposition approved the $549 million addition to NASA's FY1962 budget that was needed to get started on the accelerated program; this amount when added to the increase already approved in March represented an 89 percent increase over the previous year's budget. With this initial approval in hand, NASA could begin to implement Project Apollo.

In early May 1961, when it appeared likely that President Kennedy would approve sending Americans to the Moon, NASA associate administrator Robert Seamans asked one of his senior staff members, William Fleming, to put together a task force to examine "in detail a feasible and complete approach to the accomplishment of an early manned lunar mission." The task force considered only one approach to the lunar mission, the "direct ascent" mode, in which the very large Nova launch vehicle would send a complete spacecraft to the lunar surface. This approach had been the basis of NASA's early planning for a lunar landing. But Seamans also recognized that there were other approaches to the lunar landing that would involve rendezvous between two or more elements of a lunar spacecraft. So on May 25, the same day as President Kennedy announced the lunar landing goal, Seamans asked Bruce Lundin of NASA's Lewis Research Center to head up another group that would examine various rendezvous approaches as a way of getting to the Moon.

Lundin and his associates conducted a rapid assessment and reported back to Seamans on June 10. The group examined four rendezvous concepts: 1) rendezvous in earth orbit; 2) rendezvous in lunar orbit after takeoff from the lunar surface; 3) rendezvous in both earth and lunar orbit; 4) rendezvous on the lunar surface. They concluded that "of the various orbital operations considered, the use of rendezvous in earth orbit by two or three Saturn C-3 vehicles (depending on estimated payload requirements) was strongly favored." This approach was either the first or second choice of all members of the group.

Based on this conclusion, Seamans formed yet another group, this one to examine rendezvous approaches in more depth than had been possible in the rapid Lundin study. This group was headed by Donald Heaton of NASA Headquarters. Following Lundin's report, the group considered only earth orbital rendezvous (EOR) approaches. In its late August report, the group concluded that "rendezvous offers the earliest possibility for a successful manned lunar landing."

NASA continued to consider both a direct ascent and earth orbital rendezvous approaches for the next several months. Then, on November 15, "somewhat as a voice in the wilderness," John Houbolt, a NASA engineer at the Langley Research Center, bypassed several layers of management and wrote an impassioned, highly influential nine-page letter to Seamans arguing that NASA was overlooking the best way to get to the Moon before 1970: lunar orbital rendezvous (LOR). Although not the sole originator of the LOR mode, Houbolt became its most persistent advocate. The letter became a catalyst in shifting the thinking within NASA in favor of the LOR mode.

The Saturn C-3 that Houbolt refers to in his letter was a configuration of the Saturn booster with two massive F-1 rocket engines in its first stage. The Saturn C-2 was a less powerful booster using older engines in its first stage. The Golovin Committee was a NASA-DOD group attempting to develop a national launch vehicle program.

John C. Houbolt (NASA, Langley Research Center), Letter to Dr. Robert C. Seamans Jr. (Associate Administrator, NASA), November 15, 1961

Dear Dr. Seamans:

Somewhat as a voice in the wilderness, I would like to pass on a few thoughts on matters that have been of deep concern to me over recent months. This concern may be phrased in terms of two questions: (1) If you were told that we can put men on the moon with safe return with a single C-3, its equivalent or something less, would you judge this statement with the critical skepticism that others have? . . .

Since we have had only occasional and limited contact, and because you therefore probably do not know me very well, it is conceivable that after reading this you may feel that you are dealing with a crank. Do not be afraid of this. The thoughts expressed here may not be stated in as diplomatic a fashion as they might be, or as I would normally try to do, but this is by choice and the moment is not important. The important point is that you hear the ideas directly, not after they have filtered through a score or more of other people, with the attendant risk that they may not even reach you.

MANNED LUNAR LANDING THROUGH USE OF LUNAR ORBIT RENDEZVOUS

The plan.—The first attachment [not included in this book] outlines in brief the plan by which we may accomplish a manned lunar landing through use of a lunar rendezvous, and shows a number of schemes for doing this by means of a single C-3, its equivalent, or even something less. The basic ideas of the plan were presented before various NASA people well over a year ago, and were since repeated at numerous interlaboratory meetings. A lunar landing program utilizing rendezvous concepts was even suggested back in April. Essentially, it had three basic points: (1) the establishment of an early rendezvous program involving Mercury, (2) the specific inclusion of rendezvous in Apollo developments, and (3) the accomplishment of lunar landing through use of C-2's. It was indicated then that the two C-2's could do the job, C-2 being referred to simply because NASA booster plans did not go beyond the C-2 at that time; it was mentioned, however, that with a C-3 the number of boosters required would be cut in half, specifically only one.

Regrettably, there was little interest shown in the idea—indeed, if any, it was negative.

Also (for the record), the scheme was presented before the Lundin Committee. It received only bare mention in the final report and was not discussed further (see comments below in section entitled "Grandiose Plans").

It was presented before the Heaton Committee, accepted as a good idea, then dropped, mainly on the irrelevant basis that it did not conform to the ground rules. I even argued against presenting the main plan considered by the Heaton Committee, largely because it would only bring harm to the rendezvous cause, and further argued that if the committee did not want to consider lunar rendezvous, at least they should make a strong recommendation that it looks promising enough that it deserves a separate treatment by itself—but to no avail. In fact, it was mentioned that if I felt sufficiently strong about the matter, I should make a minority report. This is essentially what I am doing.

We have given the plan to the presently meeting Golovin Committee on several occasions.

In a rehearsal of a talk on rendezvous for the recent Apollo Conference, I gave a brief reference to the plan, indicating the benefit derivable therefrom, knowing full well that the reviewing committee would ask me to withdraw any reference to this idea. As expected, this was the only item I was asked to delete.

The plan has been presented to the Space Task Group personnel several times, dating back to more than a year ago. The interest expressed has been completely negative.

<u>**Ground rules**</u>.—The greatest objection that has been raised about our lunar rendezvous plan is that it does not conform to the "ground rules." This to me is nonsense; the important question is, "Do we want to get to the moon or not?", and, if so, why do we have to restrict our thinking along a certain narrow channel. I feel very fortunate that I do not have to confine my thinking to arbitrarily set up ground rules which only serve to constrain and preclude possible equally good or perhaps better approaches. Too often thinking goes along the following vein: ground rules are set up, and then the question is tacitly asked, "Now, with these ground rules what does it take, or what is necessary to do the job?" A design begins and shortly it is realized that a booster system way beyond present plans is necessary. Then a scare factor is thrown in; the proponents of the plan suddenly become afraid of the growth problem or that perhaps they haven't computed so well, and so they make the system even larger as an "insurance" that no matter what happens the booster will be large enough to meet the contingency. Somehow, the fact is completely ignored that they are now dealing with a ponderous development that goes far beyond the state-of-the-art.

Why is there not more thinking along the following lines: Thus, with this given booster, or this one, is there anything we can do to do the job? In other words, why can't we also think along the lines of deriving a plan to fit a booster, rather than derive a booster to fit a plan?

Three ground rules in particular are worthy of mention: three men, direct landing, and storable return. These are *very* restrictive requirements. If two men can do the job, and if the use of only two men allows the job to be done, then why not do it this way? If relaxing the direct requirements allows the job to be done with a C-3, then why not relax it? Further, when a hard objective look is taken at the use of storables, then it is soon realized that perhaps they aren't so desirable or advantageous after all in comparison with some other fuels.

Grandiose plans, one-sided objections, and bias.—For some inexplicable reason, everyone seems to want to avoid simple schemes. The majority always seems to be thinking in terms of grandiose plans, giving all sort of arguments for long-range plans, etc. Why is there not more thinking in the direction of developing the simplest scheme possible? Figuratively, why not go buy a Chevrolet instead of a Cadillac? Surely a Chevrolet gets one from one place to another just as well as a Cadillac, and in many respects with marked advantages.

I have been appalled at the thinking of individuals and committees on these matters. For example, comments of the following type have been made: "Houbolt has a scheme that has a 50 percent chance of getting a man to the moon, and a 1 percent chance of getting him back." This comment was made by a Headquarters individual at "high level" who never really has taken the time to hear about the scheme, never has had the scheme explained to him fully, or possible even correctly, and yet he feels free to pass judgment on the work. I am bothered by stupidity of this type being displayed by individuals who are in a position to make decisions which affect not only the NASA, but the fate of the nation as well. I have even grown to be concerned about the merits of all the committees that have been considering the problem. Because of bias, the intent of the committee is destroyed even before it starts and, further, the outcome is usually obvious from the beginning. We knew what the Fleming Committee results would be before it started. After one day it was clear what decisions the Lundin Committee would reach. After a couple days it was obvious what the main decision of the Heaton Committee would be. In connection with the Lundin Committee, I would like to cite a specific example. Considered by this committee was one of the most hair-brained ideas I have ever heard, and yet it received one first place vote. In contrast, our lunar rendezvous scheme, which I am positive is a much more workable idea, received only bare mention in a negative vein, as was mentioned earlier. Thus, committees are no better than the bias of the men composing them. We might then ask, why are men who are not competent to judge ideas, allowed to judge them?

Perhaps the substance of this section might be summarized this way. Why is NOVA, with its ponderous ideas, whether in size, manufacturing, erection, site location, etc., simply just accepted, and why is a much less grandiose scheme involving rendezvous ostracized or put on the defensive?

***

I make the above points because, as you will see, we have a very strong point to make about the possibility of coming up with a realistic schedule; the plan we offer is exceptionally clean and simple in vehicle and booster requirements relative to other plans.

***

CONCLUDING REMARKS

It is one thing to gripe, another to offer constructive criticism. Thus, in making a few final remarks, I would like to offer what I feel would be a sound integrated overall program. I think we should:

1. Get a manned rendezvous experiment going with the Mark II Mercury.
2. Firm up the engine program suggested in this letter and attachment, converting the booster to these engines as soon as possible.
3. Establish the concept of using a C-3 and lunar rendezvous to accomplish the manned lunar landing as a firm program.

Naturally, in discussing matters of the type touched upon herein, one cannot make comments without having them smack somewhat against NOVA. I want to assure you, however, I'm not trying to say NOVA should not be built. I'm simply trying to establish that our scheme deserves a parallel front-line position. As a matter of fact, because the lunar rendezvous approach is easier, quicker, less costly, requires less development, less new sites and facilities, it would appear more appropriate to say that this is the way to go, and that we will use NOVA as a follow on. Give us the go-ahead, and C-3, and we will put men on the moon in very short order—and we don't need any Houston empire to do it . . .

John Houbolt's letter catalyzed intense scrutiny on the best approach to carrying out the lunar landing mission. Over the next four months, both the Manned Spacecraft Center (MSC) in Houston and the Marshall Space Flight Center (MSFC) in Huntsville carried out detailed studies of alternative rendezvous approaches to getting to the Moon. Nova had by then lost favor as a feasible approach, mainly because it seemed to be too large a jump from the much smaller boosters NASA was using for the Mercury program. In addition, the concept of designing a single spacecraft to carry out all phases of the mission, particularly the lunar landing and the return into the Earth's atmosphere, looked increasingly difficult. During the early months of 1962, the MSC leadership became convinced that some version of the LOR approach, which involved developing two separate spacecraft, one specialized only for landing on the Moon and one for the journey to and from lunar orbit, was indeed the best way to proceed. The combined weight of the two spacecraft would allow the mission to be launched with a single Saturn C-5 (Saturn V) booster, although there was very little margin for weight growth. They shared their analyses and reasoning with their colleagues at MSFC, who were continuing to focus their efforts on various EOR approaches.

A climactic meeting was held at MSFC on June 7, 1962. For most of the day, the Marshall staff presented their positive findings on EOR. At the end of the day, MSFC director Wernher von Braun provided concluding remarks. He shocked many of his associates by announcing that he had concluded that his first-priority choice was the "Lunar Orbit Rendezvous Mode," because "we believe this program offers the highest confidence factor of successful accomplishment within this decade." With his endorsement, NASA soon adopted this approach in its planning for the lunar landing mission, a decision that was critical to Apollo's success.

The C-1 and C-5 vehicles referred to in von Braun's statement became known as the Saturn I and IB and the Saturn V. The C-8 was a configuration with eight first-stage engines that was never built. The S-IVB was the third stage of the Saturn V vehicle and the S-II its second stage. Robert Gilruth was the director of the Manned Spacecraft Center. NAA was North American Aviation, the contractor building the Apollo command and service module and the S-II and S-IVB stages of the Saturn C-5 (Saturn V) launcher. Rocketdyne was the company building the F-1 and J-2 rocket engines.

"Concluding Remarks by Dr. Wernher von Braun About Mode Selection for the Lunar Landing Program," June 7, 1962

***

Our general conclusion is that all four modes investigated are technically feasible and could be implemented with enough time and money. We have, however, arrived at a definite list of preferences in the following order:

1. Lunar Orbit Rendezvous Mode—with the strong recommendation (to make up for the limited growth potential of this mode) to initiate, simultaneously, the development of an unmanned, fully automatic, one-way C-5 logistics vehicle.
2. Earth Orbit Rendezvous Mode (Tanking Mode).
3. C-5 Direct Mode with minimum size Command Module and High Energy Return.
4. Nova or C-8 Mode.

I shall give you the reasons behind this conclusion in just one minute.

But first I would like to reiterate once more that *it is absolutely mandatory that we arrive at a definite mode decision within the next few weeks, preferably by the first of July, 1962.* We are already losing time in our over-all program as a result of a lacking mode decision . . .

If we do not make a clear-cut decision on the mode very soon, our chances of accomplishing the first lunar expedition in this decade will fade away rapidly.

I. WHY DO WE RECOMMEND LUNAR ORBIT RENDEZVOUS MODE PLUS C-5 ONE-WAY LOGISTICS VEHICLE?

a. We believe this program offers the highest confidence factor of successful accomplishment within this decade.

b. It offers an adequate performance margin. With storable propellants, both for the Service Module and Lunar Excursion Module, we should have a comfortable padding with respect to propulsion performance and weights. The performance margin could be further increased by initiation of a back-up development aimed at a High Energy Propulsion System for the Service Module and possibly the Lunar Excursion Module. Additional performance gains could be obtained if current proposals by Rocketdyne to increase the thrust and/or specific impulses of the F-1 and J-2 engines were implemented.

c. We agree with the Manned Spacecraft Center that the designs of a maneuverable hyperbolic re-entry vehicle and of a lunar landing vehicle constitute the two most critical tasks in producing a successful lunar spacecraft. A drastic separation of these two functions into two separate elements is bound to greatly simplify the development of the spacecraft system. Developmental cross-feed between results from simulated or actual landing tests, on the one hand, and re-entry tests, on the other, are minimized if no attempt is made to include the Command Module into the lunar landing process. The mechanical separation of the two functions would virtually permit completely parallel developments of the Command Module and the Lunar Excursion Module. While it may be difficult to accurately appraise this advantage in terms of months to be gained, we have no doubt whatsoever that such a procedure will indeed result in very substantial saving of time.

***

k.   We at the Marshall Space Flight Center readily admit that
     when first exposed to the proposal of the Lunar Orbit Ren-
     dezvous Mode we were a bit skeptical—particularly of the
     aspect of having the astronauts execute a complicated ren-
     dezvous maneuver at a distance of 240,000 miles from the
     earth where any rescue possibility appeared remote. In the
     meantime, however, we have spent a great deal of time and
     effort studying the four modes, and we have come to the
     conclusion that this particular disadvantage is far out-
     weighed by the advantages listed above.

         We understand that the Manned Spacecraft Center was
     also quite skeptical at first when John Houbolt of Langley
     advanced the proposal of the Lunar Orbit Rendezvous
     Mode, and that it took them quite a while to substantiate
     the feasibility of the method and finally endorse it.

         Against this background it can, therefore, be concluded
     that the issue of "invented here" versus "not invented here"
     does not apply to either the Manned Spacecraft Center or
     the Marshall Space Flight Center; that both Centers have
     actually embraced a scheme suggested by a third. Undoubt-
     edly, personnel of MSC and MSFC have by now conducted
     more detailed studies on all aspects of the four modes than
     any other group. Moreover, it is these two Centers to which
     the Office of Manned Space Flight would ultimately have to
     look to "deliver the goods". I consider it fortunate indeed
     for the Manned Lunar Landing Program that both Centers,
     after much soul searching, have come to identical conclu-
     sions. This should give the Office of Manned Space Flight
     some additional assurance that our recommendations
     should not be too far from the truth.

                              * * *

*As the mobilization of resources needed for Project Apollo
gained momentum, President Kennedy decided to make a personal inspection of the project's progress. On September 11,
1962, he flew to Marshall Space Flight Center in Huntsville, Alabama, where he witnessed a test firing of the first stage of a Saturn I booster and stood by as NASA administrator James Webb
and presidential science adviser Jerome Wiesner argued about
the lunar orbit rendezvous approach NASA had chosen for the*

lunar landing mission. He then flew to NASA's Launch Opera-
tion Center at Cape Canaveral, Florida, and ended the day in
Houston, Texas, where on September 12 he visited the new
Manned Spacecraft Center. As a highlight of his trip, on a hot
and humid September 12 morning, Kennedy spoke before a large
crowd assembled at Rice University's football stadium. This
speech was his fullest exposition of the rationale behind his deci-
sion to send American astronauts to the Moon. It is sometimes
confused with Kennedy's May 25, 1961, speech to Congress as
the occasion when Kennedy announced his decision to go to the
Moon.

John F. Kennedy, Address at Rice University on the Nation's Space Effort, September 12, 1962

***

The exploration of space will go ahead, whether we join in it or not, and it is one of the great adventures of all time, and no nation which expects to be the leader of other nations can expect to stay behind in this race for space.

Those who came before us made certain that this country rode the first waves of the industrial revolutions, the first waves of modern invention, and the first wave of nuclear power, and this generation does not intend to founder in the backwash of the coming age of space. We mean to be a part of it—we mean to lead it. For the eyes of the world now look into space, to the moon and to the planets beyond, and we have vowed that we shall not see it governed by a hostile flag of conquest, but by a banner of freedom and peace. We have vowed that we shall not see space filled with weapons of mass destruction, but with instruments of knowledge and understanding.

Yet the vows of this Nation can only be fulfilled if we in this Nation are first, and, therefore, we intend to be first. In short, our leadership in science and in industry, our hopes for peace and security, our obligations to ourselves as well as others, all require us to make this effort, to solve these mysteries, to solve them for the good of all men, and to become the world's leading spacefaring nation.

We set sail on this new sea because there is new knowledge to be gained, and new rights to be won, and they must be won and used for the progress of all people. For space science, like nuclear science and all technology, has no conscience of its own. Whether it will become a force for good or ill depends on man, and only if the United States occupies a position of pre-eminence can we help decide whether this new ocean will be a sea of peace or a new terrifying theater of war. I do not say that we should or will go unprotected against the hostile misuse of space any more than we go unprotected against the hostile use of land or sea, but I do say that space can be explored and mastered without feeding the fires of war, without repeating the mistakes that man has made in extending his writ around this globe of ours.

There is no strife, no prejudice, no national conflict in outer space as yet. Its hazards are hostile to us all. Its conquest deserves the best of all mankind, and its opportunity for peaceful cooperation may never come again. But why, some say, the moon? Why choose this as our goal? And they may well ask why climb the highest mountain. Why, 35 years ago, fly the Atlantic? Why does Rice play Texas?

We choose to go to the moon. We choose to go to the moon in this decade and do the other things, not because they are easy, but because they are hard, because that goal will serve to organize and measure the best of our energies and skills, because that challenge is one that we are willing to accept, one we are unwilling to postpone, and one which we intend to win, and the others, too.

It is for these reasons that I regard the decision last year to shift our efforts in space from low to high gear as among the most important decisions that will be made during my incumbency in the Office of the Presidency.

In the last 24 hours we have seen facilities now being created for the greatest and most complex exploration in man's history. We have felt the ground shake and the air shattered by the testing of a Saturn C-1 booster rocket, many times as powerful as the Atlas which launched John Glenn, generating power equivalent to 10,000 automobiles with their accelerators on the floor. We have seen the site where five F-1 rocket engines, each one as powerful as all eight engines of the Saturn combined, will be clustered together to make the advanced Saturn missile, assembled in a new building to be built at Cape Canaveral as tall as a 48-story structure, as wide as a city block, and as long as two lengths of this field.

Within these last 19 months at least 45 satellites have circled the earth. Some 40 of them were "made in the United States of America" and they were far more sophisticated and supplied far more knowledge to the people of the world than those of the Soviet Union.

The Mariner spacecraft now on its way to Venus is the most intricate instrument in the history of space science. The accuracy of that shot is comparable to firing a missile from Cape Canaveral and dropping it in this stadium between the 40-yard lines.

Transit satellites are helping our ships at sea to steer a safer course. Tiros satellites have given us unprecedented warnings of hurricanes and storms, and will do the same for forest fires and icebergs.

We have had our failures, but so have others, even if they do not admit them. And they may be less public.

To be sure, we are behind, and will be behind for some time in manned flight. But we do not intend to stay behind, and in this decade we shall make up and move ahead.

***

To be sure, all this costs us all a good deal of money. This year's space budget is three times what it was in January 1961, and it is greater than the space budget of the previous 8 years combined. That budget now stands at $5,400 million a year—a staggering sum, though somewhat less than we pay for cigarettes and cigars every year. Space expenditures will soon rise some more, from 40 cents per person per week to more than 50 cents a week for every man, woman, and child in the United States, for we have given this program a high national priority—even though I realize that this is in some measure an act of faith and vision, for we do not now know what benefits await us. But if I were to say, my fellow citizens, that we shall send to the moon, 240,000 miles away from the control station in Houston, a giant rocket more than 300 feet tall, the length of this football field, made of new metal alloys, some of which have not yet been invented, capable of standing heat and stresses several times more than have ever been experienced, fitted together with a precision better than the finest watch, carrying all the equipment needed for propulsion, guidance, control, communications, food and survival, on an untried mission, to an unknown celestial body, and then return it safely to earth, reentering the atmosphere at speeds of over 25,000 miles per hour, causing heat about half that of the temperature of the sun—almost as hot as it is here today—and do all this, and do it right, and do it first before this decade is out, then we must be bold . . .

However, I think we're going to do it, and I think that we must pay what needs to be paid. I don't think we ought to waste any money, but I think we ought to do the job. And this will be done in the decade of the sixties. It may be done while some of you are still here at school at this college and university. It will be done during the terms of office of some of the people who sit here on this platform. But it will be done. And it will be done before the end of this decade.

I am delighted that this university is playing a part in putting a man on the moon as part of a great national effort of the United States of America.

Many years ago the great British explorer George Mallory, who was to die on Mount Everest, was asked why did he want to climb it. He said, "Because it is there."

Well, space is there, and we're going to climb it, and the moon and the planets are there, and new hopes for knowledge and peace are there. And, therefore, as we set sail we ask God's blessing on the most hazardous and dangerous and greatest adventure on which man has ever embarked.

During his September 11–12 visit to the three NASA installations most involved in Project Apollo, there were suggestions made to President Kennedy (apparently by manned spaceflight head D. Brainerd Holmes) that the first lunar landing, at that point tentatively scheduled for late 1967, might actually be accomplished up to a year earlier if additional funds were provided to the Apollo program. Holmes and NASA administrator James Webb had disagreed on the wisdom of seeking additional funds for Apollo from Congress. Tensions between Holmes and Webb had been festering since at least August 1962, when Holmes was featured on the cover of Time *magazine and labeled "Apollo czar." Another* Time *story appeared on November 19, this time suggesting that Webb was not fully supporting the president's lunar landing goal and that the program was in trouble and badly needed the extra funds. Holmes was the apparent source of the story.*

The White House called a November 21 meeting in the Cabinet Room to try to understand exactly what was going on at NASA. A transcript of this secretly recorded meeting provides a rare insight into the interactions between Kennedy and Webb. During the meeting, Kennedy made an often-quoted statement that "I am not that interested in space." This transcript shows the context for that statement; Kennedy was referring primarily to the results of the space science program, not to the importance of the space program overall.

Transcript of Presidential Meeting in the Cabinet Room of the White House, November 21, 1962

***

President Kennedy: Now, let me just get back to this, what is your . . . your view is we ought to spend this four hundred forty million?

Brainerd Holmes: My view is that . . . it would accelerate the Apollo schedule, yes, sir. Let me say I was very . . . I ought to add that I'm very sorry about this . . . I have no disagreement with Mr. Webb . . . I think my job is to say how fast I think we can go for what dollars.

James Webb: Well, I think it's fair to say one other thing, Mr. President, that after your visit when you were saying how close this was, the speech you made. I think Brainerd and Wernher von Braun and Gilruth all felt, "We've got to find out how fast we can move here. The President wants to move." So they went to the contractors and said, "How fast can you move, boys, if money were not a limit?" Now, this sort of got cranked up into a feeling that this money was going to be made available, that a policy decision had already been made to ask for the supplemental. And I think, to a certain extent, then, the magazines like *Time*, they picked this up in order to make a controversy.

***

James Webb: Well, let me make a statement on that I have made to the Budget Director. You remember when I first talked to you about this program, the first statement I made to Congress was that the lunar program would cost between twenty and forty billion dollars. Now I am able to say right now it's going to be under the twenty billion, under the lower limit that we used. The question is how rapidly do you spend the money and . . . and how efficiently you manage this so as to get the most possible for the money. This can be speeded up at the expense of . . . of certain things which I outlined in this letter to you. It can be slowed up if, a year from now, we find that we don't have to proceed at this basis. But this is a good, sound, solid program that would keep all of the governmental agencies and the contractors and the rest moving ahead. But we're prepared to move if you really want to put it on a crash basis.

President Kennedy: Do you think this program is the top-priority of the Agency?

James Webb: No, sir, I do not. I think it is *one* of the top-priority programs, but I think it's very important to recognize here . . . and that you have found what you could do with a rocket as you could find how you could get out beyond the Earth's atmosphere and into space and make measurements. Several scientific disciplines that are the very powerful and beginning to converge on this area.

President Kennedy: Jim, I think it is the top priority. I think we ought to have that very clear. Some of these other programs can slip six months, or nine months, and nothing strategic is going to happen . . . But this is important for political reasons, international political reasons. This is, whether we like it or not, in a sense a race. If we get second to the Moon, it's nice, but it's like being second any time. So that if we're second by six months, because we didn't give it the kind of priority, then of course that would be very serious. So I think we have to take the view that this is the top priority with us.

James Webb: But the environment of space is where you are going to operate Apollo and where you are going to do the landing.

President Kennedy: Look, I know all these other things and the satellite and the communications and weather and all, they're all desirable, but they can wait.

James Webb: I'm not putting those . . . I am talking now about the scientific program to understand the space environment within which you've got to fly Apollo and make a landing on the Moon.

President Kennedy: Wait a minute—is that saying that the lunar program to land the man on the Moon is the top priority of the Agency, is it?

Unknown speaker: And the science that goes with it.

Robert Seamans: Well, yes, if you add that, the science that is necessary.

President Kennedy: The science . . . Going to the Moon is the top-priority project. Now, there are a lot of related scientific information and developments that will come from that which are important. But the whole thrust of the Agency, in my opinion, is the lunar program. The rest of it can wait six or nine months.

James Webb: . . . Let me say one thing . . . the thing that troubles me here about making such a flat statement as that is, number one, there are real unknowns as to whether man can live under the weightless condition and you'll ever make the lunar landing. This is one kind of political vulnerability I'd like to avoid such a flat commitment to. If you say you failed on your number-one priority, this is something to think about. Now, the second point is that as we can go out and make measurements in space by being physically able to get there, the scientific work feeds the technology and the engineers begin to make better spacecraft. That gives you better instruments and a better chance to go out to learn more. Now right all through our universities some of the brilliant able scientists are recognizing this and beginning to get into this area and you are generating here on a national basis an intellectual effort of the highest order of magnitude that I've seen develop in this country in the years I've been fooling around with national policy. Now, to them, there is a real question. The people that are going to furnish the brainwork, the real brainwork, on which the future space power of this nation for twenty-five or a hundred years are going be to made, have got some doubts about it . . .

President Kennedy: Doubts about what, with this program?

James Webb: As to whether the actual landing on the Moon is what you call the highest priority.

President Kennedy: What do they think is the highest priority?

James Webb: They think the highest priority is to understand the environment and . . . and the areas of the laws of nature that operate out there as they apply backwards into space. You can say it this way. I think Jerry [Wiesner] ought to talk on this rather than me, but the scientists in the nuclear field have penetrated right into the most minute areas of the nucleus and the subparticles of the nucleus. Now here, out in the universe, you've got the same general kind of a structure, but you can do it on a massive universal scale.

President Kennedy: I agree that we're interested in this, but we can wait six months on all of it.

James Webb: But you have to use that information to . . .

President Kennedy: Yes, but only as that information directly applies to the program. Jim, I think we've got to have that . . .

Jerome Wiesner: Mr. President, I don't think Jim understands some of the scientific problems that are associated with landing on the Moon and this is what Dave Bell was trying to say and what I'm trying to say. We don't know a damn thing about the surface of the Moon. And we're making the wildest guesses about how we're going to land on the Moon and we could get a terrible disaster from putting something down on the surface of the Moon that's very different than we think it is. And the scientific programs that find us that information have to have the highest priority. But they are associated with the lunar program. The scientific programs that aren't associated with the lunar program can have any priority we please to give them.

Unknown speaker: That's consistent with what the President was saying.

Robert Seamans: Could I just say that I agree with what you say, Jerry, that we must gather a wide variety of scientific data in order to carry out the lunar mission. For example, we must know what conditions we'll find on the lunar surface. That's the reason that we are proceeding with Centaur in order to get the Surveyor unmanned spacecraft to the Moon in time that it could affect the design of the Apollo.

President Kennedy: The other thing is I would certainly not favor spending six or seven billion dollars to find out about space no matter how on the schedule we're doing. I would spread it out over a five- or ten-year period . . . Why are we spending seven million dollars on getting fresh water from saltwater, when we're spending seven billion dollars to find out about space? Obviously, you wouldn't put it on that priority except for the defense implications. And the second point is the fact that the Soviet Union has made this a test of the system. So that's why we're doing it. So I think we've got to take the view that this is the key program. The rest of this . . . we can find out all about it, but there's a lot of things we can find out about; we need to find out about cancer and everything else . . . Everything that we do ought to really be tied into getting onto the Moon ahead of the Russians.

James Webb: Why can't it be tied to preeminence in space . . .

President Kennedy: Because, by God, we keep, we've been telling everybody we're preeminent in space for five years and nobody believes it because they have the booster and the satellite. We know all about the number of satellites we put up, two or three times the number of the Soviet Union . . . We're ahead scientifically. It's like that instrument you got up at Stanford which is costing us a hundred and twenty-five million dollars and everybody tells me that we're the number one in the world. And what is it? I can't think what it is.

Interruption from multiple unknown speakers: The linear accelerator.

President Kennedy: I'm sorry, that's wonderful, but nobody knows anything about it!

James Webb: Let me say it slightly different. The advanced Saturn is eighty-five times as powerful as the Atlas. Now we are building a tremendous giant rocket with an index number of eighty-five if you give Atlas one. Now, the Russians have had a booster that will lift fourteen thousand pounds into orbit. They've been very efficient and capable with it. The kinds of things I'm talking about that give you preeminence in space are what permit you to make either that Russian booster or the advanced Saturn better than any other. A range of progress is possible . . .

President Kennedy: . . . We're not going to settle the four hundred million this morning. I want to take a look closely at what Dave Bell . . . But I do think we ought [to] get it, you know, really clear that the policy ought to be that this is *the* top-priority program of the Agency, and one of the two things, except for defense, the top priority of the United States government. I think that that is the position we ought to take. Now, this may not change anything about that schedule, but at least we ought to be clear, otherwise we shouldn't be spending this kind of money because I'm not that interested in space. I think it's good; I think we ought to know about it; we're ready to spend reasonable amounts of money. But we're talking about these *fantastic* expenditures which wreck our budget and all these other domestic programs and the only justification for it, in my opinion, to do it in this time or fashion, is because we hope to beat them and demonstrate that starting behind, as we did by a couple years, by God, we passed them.

James Webb: I'd like to have more time to talk about that because there is a wide public sentiment coming along in this country for preeminence in space.

President Kennedy: If you're trying to prove preeminence, this is the way to prove your preeminence . . . We do have to talk about this. Because I think if this affects in any way our sort of allocation of resources and all the rest, then it is a substantive question and I think we've got to get it clarified. I'd like to have you tell me in a brief . . . You write me a letter, your views. I'm not sure that we're far apart. I think all these programs which contribute to the lunar program are . . . come within, or contribute significantly or really in a sense . . . Let's put it this way, are *essential*, put it that way . . . *are essential* to the success of the lunar program, are justified. Those that are not essential to the lunar program, that help contribute over a broad spectrum to our preeminence in space, are secondary. That's my feeling.

***

When he assumed the presidency in January 1961, Kennedy hoped that outer space could be an arena for cooperation

*between the United States and the Soviet Union. The domestic
and international reaction to the April 12, 1961, flight of Yuri
Gagarin convinced Kennedy that the United States had to match,
if not excel, the Soviet Union in space achievement before balanced cooperation might be possible. Even so, he proposed to
Soviet leader Nikita Khrushchev at their only summit meeting,
in Vienna, Austria, on June 3–4, 1961, that the United States
and the Soviet Union cooperate in going to the Moon. Khrushchev's response was that such cooperation would not be possible
without comprehensive disarmament. In 1963, after the peaceful resolution of the October 1962 Cuban Missile Crisis changed
the political climate between the two superpowers and after the
United States showed that it was serious about winning the race
to the Moon, Kennedy judged that it was time to raise again the
cooperative possibility. Rather than sending feelers through diplomatic channels to determine whether the Soviet Union had
changed its position regarding cooperation, Kennedy chose to
make his suggestion in a most public fashion when he addressed
the United Nations General Assembly as it opened its eighteenth
session. His offer was part of a longer speech calling for a variety
of initiatives to lower the risk of armed conflict.*

**John F. Kennedy, "Address Before the 18th General Assembly
of the United Nations," September 20, 1963**

* * *

Finally, in a field where the United States and the Soviet Union
have a special capacity—in the field of space—there is room for
new cooperation, for further joint efforts in the regulation and
exploration of space. I include among these possibilities a joint
expedition to the moon. Space offers no problems of sovereignty;
by resolution of this Assembly, the members of the United Nations have foresworn any claim to territorial rights in outer space
or on celestial bodies, and declared that international law and
the United Nations Charter will apply. Why, therefore, should
man's first flight to the moon be a matter of national competition? Why should the United States and the Soviet Union, in preparing for such expeditions, become involved in immense
duplications of research, construction, and expenditure? Surely

we should explore whether the scientists and astronauts of our two countries—indeed of all the world—cannot work together in the conquest of space, sending some day in this decade to the moon not the representatives of a single nation, but the representatives of all of our countries.

**

Kennedy's proposal was greeted with dismay by many of those who had been Apollo's strongest supporters. For example, Congressman Albert Thomas, who chaired the House Appropriations Subcommittee that controlled the NASA budget, sent a handwritten note to the president the day after the speech, saying that "the press and many private individuals seized upon your offer to cooperate with the Russians in a moon shot as a weakening of your former position of a forthright and strong effort in lunar landings." Thomas asked the president for "a letter clarifying your position with reference to our immediate effort in this regard."

Kennedy replied to Thomas on September 23, explaining how his proposal was consistent with a strong Apollo effort.

Letter from John F. Kennedy to Representative Albert Thomas, September 23, 1963

Dear Al:

I am very glad to respond to your letter of September 21 and to state my position on the relation between our great current space effort and my proposal at the United Nations for increased cooperation with the Russians in this field. In my view an energetic continuation of our strong space effort is essential, and the need for this effort is, if anything, increased by our intent to work for increasing cooperation if the Soviet Government proves willing.

As you know, the idea of cooperation in space is not new. My statement of our willingness to cooperate in a moon shot was an extension of a policy developed as long ago as 1958 on a bipartisan basis, with particular leadership from Vice President Johnson, who was then the Senate Majority Leader. The American purpose of cooperation in space was stated by the Congress in the National Aeronautics and Space Act of 1958, and reaffirmed in my Inaugural Address in 1961. Our specific interest in cooperation with the Soviet Union, as the other nation with a major present capability in space, was indicated to me by Chairman Khrushchev in Vienna in the middle of 1961, and reaffirmed in my letter to him of March 7, 1962, which was made public at the time. As I then said, discussion of cooperation would undoubtedly show us "possibilities for substantive scientific and technical cooperation in manned and unmanned space investigations." So my statement in the United Nations is a direct development of policy long held by the United States government.

Our repeated efforts of cooperation with the Soviet Union have so far produced only limited responses and results. We have an agreement to exchange certain information in such limited fields as weather observation and passive communications, and technical discussions of other limited possibilities are going forward. But as I said in July of this year, there are a good many barriers of suspicion and fear to be broken down before we can have major progress in this field. Yet our intent remains: to do our part to bring those barriers down.

At the same time, as no one knows better than you, the United States in the last five years has made a steadily growing national effort in space. On May 25, 1961, I proposed to the Congress and the nation a major expansion of this effort, and I particularly emphasized as a target the achievement of a manned lunar landing in the decade of the 60's. I stated that this would be a task requiring great effort and very large expenditures. The Congress and the nation approved this goal; we have been on our way ever since. In a larger sense this is not merely an effort to put a man on the moon; it is a means and a stimulus for all the advances in technology, in understanding and in experience, which can move us forward toward man's mastery of space.

This great national effort and this steadily stated readiness to cooperate with others are not in conflict. They are mutually supporting elements of a single policy. We do not make our space effort with the narrow purpose of national aggrandizement. We make it so that the United States may have a leading and honorable role in mankind's peaceful conquest of space. It is this great effort which permits us now to offer increased cooperation with no suspicion anywhere that we speak from weakness. And in the same way, our readiness to cooperate with others enlarged the international meaning of our own peaceful American program in space.

In my judgment, therefore, our renewed and extended purpose of cooperation, so far from offering any excuse for slackening or weakness in our space effort, is one reason the more for moving ahead with the great program to which we have been committed as a country for more than two years.

So the position of the United States is clear. If cooperation is possible, we mean to cooperate, and we shall do so from a position made strong and solid by our national effort in space. If cooperation is not possible—and as realists we must plan for this contingency too—then the same strong national effort will serve all free men's interest in space, and protect us also against possible hazards to our national security. So let us press on.

Let me thank you again for this opportunity of expressing my views.

The idea of U.S.-Soviet cooperation in going to the Moon was very much a Kennedy personal initiative. Momentum behind the cooperative proposal dissipated after Kennedy's assassination; in addition, the Soviet Union never responded to Kennedy's September 20 offer.

On June 12, 1963, with relations between James Webb and Brainerd Holmes still tense, Holmes submitted his resignation. This meant that Apollo was losing the leader who in the eyes of the public and media had come to personify the effort. It took NASA a little more than a month to settle on a replacement for Holmes. The individual selected, George Mueller, was vice president for research and development of the company Space Technology Laboratories, which was deeply involved in Air Force missile programs; Mueller would report to NASA on September 1. Unlike Holmes, who courted media attention, Mueller focused his attention on relationships between NASA Headquarters, the NASA field centers, NASA's contractors, and the Congress. One account of the Apollo program describes Mueller as "brilliant," "intellectually arrogant," and "a complex man." Robert Seamans characterized him as "tireless."

In the next several months Mueller made a number of key

personnel changes. He and George Low did not get along, and Low moved to the Manned Spacecraft Center in Houston as deputy director under Robert Gilruth. On December 31, Air Force Brigadier General Samuel Phillips took over the Apollo Program Office at NASA Headquarters. The team of Mueller and Phillips was to provide strong leadership as the Apollo program encountered both tragedy and triumph.

Soon after he came to NASA, Mueller asked two veteran NASA engineers not directly involved in Apollo, John Disher and Del Tischler, to discreetly conduct an independent assessment of the situation within Apollo. They reported to Mueller on September 28 with the troubling conclusions that the "lunar landing cannot likely be attained within the decade with acceptable risk" and that the "first attempt to land men on moon is likely about late 1971."

Bold steps were needed to get Apollo on a schedule that had a good chance of meeting President Kennedy's goal of a lunar landing before 1970, and Mueller soon took them. First he canceled flights of the Saturn I booster so that attention could be shifted to the more powerful Saturn IB, which would use the same upper stage as the Saturn V. At an October 29 meeting of his Management Council, with the senior leadership from Houston and Huntsville present, Mueller announced a new approach to getting ready for missions to the Moon that soon became known as "all-up testing." In this approach, all stages of the three-stage Saturn rocket would be tested together, rather than testing each stage on a separate flight. Two days later Mueller sent a teletype message to the Apollo field centers proposing a new, accelerated schedule of Apollo flights; in this message, he reiterated his "desire that 'all-up' spacecraft and launch vehicle flights be made as early as possible in the program. To this end, SA-201 [the first flight of the Saturn IB] and 501 [the first flight of the Saturn V] should utilize all live stages and should carry complete spacecraft for their respective missions."

Details of the "all-up" approach were provided in an October 31, 1963, memorandum to the directors of the NASA centers involved in Apollo. Von Braun and his staff at the Marshall Space Flight Center were "incredulous" when they first heard of Mueller's dictate. It violated the step-by-step approach to rocket *testing they had been following since their time in Germany. But they could not provide compelling counterarguments, particularly given the pressure to have the first lunar landing attempt come before the end of 1969. Von Braun was later to agree: "In retrospect it is clear that without all-up testing the first manned lunar landing could not have taken place as early as 1969." Mueller's "all-up" decision thus joined the selection of lunar orbit rendezvous as keys to Apollo's success.*

George E. Mueller (Deputy Associate Administrator for Manned Space Flight, NASA), to the Directors of the Manned Spacecraft Center, Launch Operations Center, and Marshall Space Flight Center, "Revised Manned Space Flight Schedule," October 31, 1963

Recent schedule and budget reviews have resulted in a deletion of the Saturn I manned flight program and realignment of schedules and flight mission assignments on the Saturn IB and Saturn V programs. It is my desire at this time to plan a flight schedule which has a good probability of being met or exceeded.

***

It is my desire that "all-up" spacecraft and launch vehicle flights be made as early as possible in the program. To this end, SA-201 and 501 should utilize all live stages and should carry complete spacecraft for their respective missions. SA-501 and 502 missions should be reentry tests of the spacecraft at lunar return velocity. It is recognized that the Saturn IB flights will have CM/SM [Command Module/Service Module] and CM/SM/LEM [Command Module/Service Module/Lunar Excursion Module] configurations.

Mission planning should consider that two successful [Saturn IB] flights would be made prior to a manned flight. Thus, 203 could conceivably be the first manned Apollo flight. However, the official schedule would show the first manned flight as 207, with flights 203–206 designated as "man-rating" flights. A similar philosophy would apply to Saturn V for "man-rating" flights with 507 shown as the first manned flight.

I would like your assessment of the proposed schedule, including any effect on resource requirements in FY 1964, 1965 and run-out by November 11, 1963. My goal is to have an official schedule reflecting the philosophy outlined here by November 25, 1963.

Though Cold War political aims were the driving factor behind Apollo, the project did have some limited scientific objectives. This memorandum reflects the advice of the Space Science Board of the National Academy of Sciences, which had addressed scientific priorities for Project Apollo. Astronaut stay time on the lunar surface would be relatively short, ranging from a few hours to a few days; this limited scientific ambitions.

Memorandum from Verne C. Fryklund Jr. (Acting Director, Manned Space Sciences Division, Office of Space Sciences, NASA Headquarters), to Robert R. Gilruth (Director, Manned Spacecraft Center), "Scientific Guidelines for the Project Apollo," October 8, 1963

***

The Office of Space Sciences has established that the primary scientific objective of the Apollo project is acquisition of comprehensive data about the Moon . . . As the moon itself is the primary subject of observation, it follows that the structure of the moon's surface, gross body properties and large-scale measurements of physical and chemical characteristics, and observation of whatever phenomena may occur at the actual surface will be prime scientific objectives.

The guidelines that follow are intended to place some specific constraints on studies in keeping with the paragraphs above.

Guidelines:

1. The principal scientific activity will be observation of the moon.

2. The use of the moon as a platform for making astronomical and other observations is, in general, not a function of the Apollo project . . .

3. We may assume that Apollo activities will be largely reconnaissance in nature. The intention is to acquire knowledge of as large an area as possible, and by as simple a means as possible, in the limited time available.

***

8.   Sample collecting, for geological and biological purposes, will be an important activity and possible special equipment requirements should be studied.

President John F. Kennedy was assassinated on November 22, 1963. A few months afterward, Kennedy's top adviser, Theodore Sorensen, was interviewed by Carl Kaysen, another close Kennedy associate. Sorensen provides a fascinating insider's view of President Kennedy's attitude toward space competition and cooperation.

John F. Kennedy Presidential Library, "Excerpts from an Oral History Interview with Theodore Sorensen," March 26, 1964

KAYSEN: Ted, I want to begin by asking you about something on which the President expressed himself very strongly in the campaign and early in his Administration, and that is space. What significance, in your mind, did the President attach to the space race in terms of, one, competition with the Soviet Union and, two, the task which the United States ought to do whether or not the element of competition with the Soviet Union was important in it?

SORENSEN: It seems to me that he thought of space primarily in symbolic terms. By that I mean he had comparatively little interest in the substantive gains to made from this kind of scientific inquiry. He did not care as much about new breakthroughs in space medicine or planetary exploration as he did new breakthroughs in rocket thrust or humans in orbit. Our lagging space effort was symbolic, he thought, of everything of which he complained in the Eisenhower Administration: the lack of effort, the lack of initiative, the lack of imagination, vitality, and vision; and the more the Russians gained in space during the last few years in the fifties, the more he thought it showed up the Eisenhower Administration's lag in this area and damaged the prestige of the United States abroad.

KAYSEN: So that your emphasis was on general competitiveness but not specific competitiveness with the Soviet Union in a military sense. The President never thought that the question of who was first in space was a big security issue in any direct sense.

SORENSEN: That's correct.

KAYSEN: Now the first big speech and the first big action on space was taken in a special message on extraordinary needs to the Congress in May. What accounted for this delay? What was the President doing in the period between his inauguration and May? He didn't really say much about space in the State of the Union message. He mentioned the competition with the Soviet Union in his State of the Union message, but he didn't really say much or present any programs. What was going on in this period between the inauguration and the inclusion of space in a message which was devoted to extraordinary, urgent was the word, urgent national need?

SORENSEN: There was actually a considerable step-up in our space effort in the first space supplementary budget which he sent to the Congress . . . My recollection is that it emphasized more funds for the Saturn booster. Then came the first Soviet to orbit the earth—[Yuri] Gagarin, I believe that was—and the President felt, justifiably so, that the Soviets had scored a tremendous propaganda victory, that it affected not only our prestige around the world, but affected our security as well in the sense that it demonstrated a Soviet rocket thrust which convinced many people that the Soviet Union was ahead of the United States militarily. First we had a very brief inquiry—largely because the President was being interviewed by Hugh Sidey of *Time* magazine and wanted to be prepared to say where we stood, what we were going to do, what we were unable to do, how much it would cost and so on—in which he asked me and [Jerome B.] Wiesner and others to look into our effort in some detail.

I do not remember the exact time sequence, but I believe it was shortly after that he asked the Vice President, as the chairman of the Space Council, to examine and to come up with the answers to four or five questions of a similar nature: What were we doing that was not enough? What more could we be doing? Where should we be trying to compete and get ahead? What would we have to do to get ahead? and so on. That inquiry led to a joint study by the Space Administration and the Department of Defense. Inasmuch as that study was going on

simultaneously with the studies and reviews we were making
of the defense budget, military assistance, and civil defense,
and inasmuch as space, like these other items, obviously did
have some bearing upon our status in the world, it was decided to combine the results of all those studies with the President's recommendations in the special message to Congress.

KAYSEN: Was the moon goal chosen as the goal for the space
program because it was spectacular, because it was the first
well-defined thing which the experts thought we could sensibly say we ought to pick as a goal we could be first in, because
it was far enough away so that we could have a good chance of
being first? What reason did we have for defining this as the
goal of the space program and making it the center of the
space element of the message?

SORENSEN: The scientists listed for us what they considered to
be the next series of steps to be taken in the exploration of
space which any major country would take, either the Soviet
Union or the United States. They included manned orbit, two
men in orbit, laboratory in orbit, a shot around the moon, a
landing of instruments on the moon, etc. In that list, then,
came the sending of a man, or a team of men, to the moon and
bringing them back safely. After that was exploration of the
planets and so forth.

   Looking at that list, the scientists were convinced—on the
basis of what they assumed to be the Russian lead at that
time—that with respect to all of the items on the list between
where we were then, in early 1961, and the landing of a man
on the moon, sometime in the late 1960's or early 1970's, there
was no possibility of our catching up with the Russians. There
was a possibility, if we put enough effort into it, of being the
first to send a team to the moon and bringing it back. And it
was decided to focus our space effort on that objective.

KAYSEN: Now, as early as the inaugural message, the President
talked about making space an area of cooperation instead of
conflict. He repeated this notion in his speech to the U.N. September '61, although with a rather narrow set of specifics on
weather and communications satellites and things like that. At
various times in the course of '61 and '62, I think the record
suggests that there was a division of emphasis between the
competitive element with the Soviet Union and the notion of

offering to cooperate in space in the President's 1963 speech to the U.N., he made a specific suggestion that we cooperate in going to the moon. Do you think this represented a change in emphasis, do you think it represented a change in the assessment of our relations with the Soviets, or do you think it represented a change in the assessment of the feasibility and desirability of trying to meet the goal set of getting to the moon in 1970 and being the first on the moon?

SORENSEN: I don't believe it represented the latter. It may have had an element of the first two in it. I think the President had three objectives in space. One was to assure its demilitarization. The second was to prevent the field from being occupied by the Russians to the exclusion of the United States. And the third was to make certain that American scientific prestige and American scientific effort were at the top. Those three goals all would have been assured in a space effort which culminated in our beating the Russians to the moon. All three of them would have been endangered had the Russians continued to outpace us in their space effort and beat us to the moon. But I believe all three of those goals would also have been assured by a joint Soviet-American venture to the moon.

The difficulty was that in 1961, although the President favored joint effort, we had comparatively few chips to offer. Obviously the Russians were well ahead of us at that time in space exploration, at least in terms of the bigger, more dramatic efforts of which the moon shot would be the culmination. But by 1963, our effort had accelerated considerably. There was a very real chance that we were even with the Soviets in this effort. In addition, our relations with the Soviets, following the Cuban missile crisis and the test ban treaty, were much improved—so the President felt that, without diminishing our own space effort, and without harming any of those three goals, we now were in a position to ask the Soviets to join with us and make it more efficient and economical for both countries.

KAYSEN: In this last element, was the President persuaded, as some people argued, that the Soviets weren't really in the race; that, for example, we were developing the Saturn, our intelligence suggested to us that the Soviets had no development of comparable thrust and character; and that, in a sense, we were racing with ourselves, and we'd won, because once we'd make the commitment to develop the Saturn and it looked as if this was feasible, although maybe the schedule wasn't clear, that we could do it and the Soviets really didn't have anything that could match that; and that, therefore, the psychological moment had come to sort of make it clear to them that we knew it?

SORENSEN: I don't know if that was in his mind. I did not know that.

KAYSEN: Now, this is a speculative question, but do you think once an offer of cooperation that was more than trivial, that went beyond the kind of things we had agreed, about exchange of weather information or other rather minor and technical points about recovery of parts and all that kind of thing, that once any offer of cooperation of that sort was made and accepted and some cooperation actually started to take place, do you think space would have become politically uninteresting?

SORENSEN: Politically, in domestic politics?

KAYSEN: Yes.

SORENSEN: It probably would have been less interesting, that's right.

KAYSEN: I'm assuming, and I take it you're assuming, that in the initial exchanges there'd be static and the right wing of the Republicans would shout and so on, but I'm assuming we'd get past all that and some actually useful cooperation would result?

SORENSEN: I think it would be less interesting. Even though the President would stress from time to time that the idea of a race or competition was not our sole motivation, there was no doubt that that's what made it more interesting to the Congress and to the general public.

KAYSEN: Was there any indication that you are aware of that in '63, that in the process of assembling the budget for '63, at the time of the first review, midyear review—that is, I'm talking about the '65 budget, of course, which took place in '63—just the size of this program and its rate of growth were beginning to worry the President, and that he was more eager to stress the cooperative issue because he was dubious about either the wisdom or the possibility of maintaining the kind of rate of increase in this program that NASA talked about?

SORENSEN: I think he was understandably reluctant to continue that rate of increase. He wished to find ways to spend less money on the program and to cut out the fat which he was convinced was in the budget. How much that motivated his offer to the Russians, though, I don't know.

KAYSEN: What would you assign it to? You'd say that the political interest in trying to find positive things we could do together was much more important than any budgetary concern about the space program or any feeling that this was not the most important effort that ought to be maintained.

SORENSEN: Right.

With Kennedy's assassination, Apollo became a memorial to a fallen young president; new president Lyndon B. Johnson gave no thought to canceling the Moon program. From 1964 to 1966, there was significant (though troubled) progress in the program aspects of Apollo. The NASA budget peaked at $5.25 billion in FY1965 and then began a gradual decline. President Johnson found himself constrained by the budget demands of his Great Society programs and the war in Vietnam, and was unwilling to provide significant financial support for major post-Apollo space initiatives. Congress continued to question whether NASA needed all the resources it was requesting to complete Apollo, and was equally unwilling to support major new programs. Public criticism of Apollo as representing misplaced priorities was widespread.

In late 1965, Major General Samuel Phillips, Apollo program director at NASA Headquarters, initiated a review of the work of North American Aviation, Inc. (referred to in the following document as NAA) to determine why the company was behind schedule and over budget on its work on both the Apollo command and service module and the second (S-II) stage of the Saturn V launch vehicle. This highly critical review was transmitted to North American's president, Lee Atwood, on December 19, 1965. Two years later, the review took on added significance in the aftermath of the fatal Apollo 1 fire, when it was discovered that NASA administrator James E. Webb was apparently unaware of its existence; it became known as the "Phillips Report."

Letter to J. Leland Atwood (President, North American Aviation, Inc.) from Major General Samuel C. Phillips (USAF, Apollo Program Director), December 19, 1965, with attached "NASA Review Team Report"

Dear Lee:

I believe that I and the team that worked with me were able to examine the Apollo Spacecraft and S-II stage programs at your Space and Information Systems Division in sufficient detail during our recent visits to formulate a reasonably accurate assessment of the current situation concerning these two programs.

I am definitely not satisfied with the progress and outlook of either program and am convinced that the right actions now can result in substantial improvement of position in both programs in the relatively near future . . .

The conclusions expressed in our briefing and notes are critical. Even with due consideration of hopeful signs, I could not find a substantial basis for confidence in future performance. I believe that a task group drawn from NAA at large could rather quickly verify the substance of our conclusions, and might be useful to you in setting the course for improvements.

The gravity of the situation compels me to ask that you let me know, by the end of January if possible, the actions you propose to take. If I can assist in any way, please let me know.

NASA REVIEW TEAM REPORT

I. Introduction

This is the report of the NASA's Management Review of North American Aviation Corporation management of Saturn II Stage (S-II) and Command and Service Module (CSM) programs. The Review was conducted as a result of the continual failure of NAA to achieve the progress required to support the objective of the Apollo Program.

***

IV. Summary Findings

Presented below is a summary of the team's views on those program conditions and fundamental management deficiencies that are impeding program progress and that require resolution by NAA to ensure that the CSM and S-II Programs regain the required program position . . .

A. NAA performance on both programs is characterized by continued failure to meet committed schedule dates with required technical performance and within costs. There is no evidence of current improvement in NAA's management of these programs of the magnitude required to give confidence that NAA performance will improve at the rate required to meet established Apollo program objectives.

***

To the outside observer, all elements of the Apollo program appeared to be moving forward toward a lunar landing before the end of the decade, with the first flight of the Apollo spacecraft with a crew aboard scheduled for early 1967. Planning for the scientific experiments to be done on the Moon was under way, and the June 1966 landing of the Surveyor 1 spacecraft on the lunar surface indicated that the surface could support the landing of the heavier lunar module, contrary to the speculation of some scientists. During 1965 and 1966, a series of ten mostly successful Gemini launches demonstrated many of the capabilities, particularly rendezvous and docking, that would be needed for Apollo.

The technical reality was rather different. There were major problems with the Apollo spacecraft and the S-II second stage of the Saturn V launcher, both being developed by North American Aviation, and the lunar module being developed by Grumman was running well behind schedule and was overweight. By the end of 1966, Apollo's Washington managers were stressing publicly that it would be difficult to attempt an initial lunar landing mission until sometime in the second half of 1969.

Despite these concerns, there was a fair degree of optimism as 1967 began, with the first crew-carrying flight of Apollo (an Earth-orbital test mission of the Apollo command and service modules which NASA designated Apollo 204) scheduled for launch on February 21. The crew consisted of veteran astronauts Virgil "Gus" Grissom and Edward White and rookie Roger Chaffee, who identified their mission as Apollo 1.

At 1:00 p.m. on January 27, 1967, the crew was strapped into the spacecraft as it sat atop an unfueled Saturn IB launcher on Pad 34 at Cape Canaveral for a lengthy countdown test. At 6:31, as the test neared its end, Roger Chaffee exclaimed, "We've got a fire in the cockpit." Within less than a minute, the three astronauts were dead of asphyxiation as they inhaled toxic gases created by the fire within the still-sealed spacecraft.

James Webb, Robert Seamans, and George Mueller learned of the fire within a few minutes. Webb immediately notified President Johnson; later the three huddled at NASA Headquarters to decide how to proceed. They decided to try to convince the president to let NASA manage the accident investigation rather than have the White House appoint an external investigation board. They were successful. By the next day, the Apollo 204 Review Board had been named; it was chaired by Floyd Thompson, director of NASA's Langley Research Center, and had eight other members from both within and outside of NASA.

In the wake of the accident, the widows of the astronauts requested that the designation "Apollo 1" be reserved for the failed test. Apollo 2 and 3 missions were not flown; NASA designated the next mission, a launch of the Apollo spacecraft without a crew aboard, as Apollo 4.

Apollo 204 Review Board, "Report of Apollo 204 Review Board to the Administrator, National Aeronautics and Space Administration," April 5, 1967

The spacecraft (S/C) consists of a launch escape system (LES) assembly, command module (C/M), service module (S/M) and the spacecraft/lunar module adapter (SLA). The LES assembly provides the means for rapidly separating the C/M from the S/M during pad or suborbital aborts. The C/M forms the spacecraft control center, contains necessary automatic and manual equipment to control and monitor the spacecraft systems, and contains the required equipment for safely and comfort of the crew. The S/M is a cylindrical structure located between the C/M and the SLA. It contains the propulsion systems for attitude and velocity change maneuvers. Most of the consumables used in the mission are stored in the S/M. The SLA is a truncated cone which connects the S/M to the launch vehicle. It also provides the space wherein the lunar module (L/M) is carried on lunar missions.

Test in progress at time of accident

Spacecraft 012 was undergoing a "Plugs Out Integrated Test" at the time of the accident on January 27, 1967. Operational Checkout Procedure, designated OCP FO-K-0021-1, applied to this test. Within this report this procedure is often referred to as OCP-0021.

***

PART VI BOARD FINDINGS, DETERMINATIONS AND RECOMMENDATIONS

In this Review, the Board adhered to the principle that reliability of the Command Module and the entire system involved in its operation is a requirement common to both safety and mission success. Once the Command Module has left the earth's environment the occupants are totally dependent upon it for their safety. It follows that protection from fire as a hazard involves much more than quick egress. The latter has merit only during test periods on earth when the Command Module is being readied for its mission and not during the mission itself. The risk of fire must be faced; however, that risk is only one factor pertaining to the reliability of the Command Module that must receive adequate consideration. Design features and operating procedures that are intended to reduce the fire risk must not introduce other serious risks to mission success and safety.

1. FINDING:
    a. There was a momentary power failure at 23:30:55 GMT.
    b. Evidence of several arcs was found in the post fire investigation.
    c. No single ignition source of the fire was conclusively identified.

DETERMINATION:

The most probable initiator was an electrical arc in the sector between the –Y and +Z spacecraft axes. The exact location best fitting the total available information is near the floor in the lower forward section of the left-hand equipment bay where Environmental Control System (ECS) instrumentation power wiring leads into the area between the Environmental Control Unit (ECU) and the oxygen panel. No evidence was discovered that suggested sabotage.

2. FINDING:
    a. The Command Module contained many types and classes of combustible material in areas contiguous to possible ignition sources.
    b. The test was conducted with a 16.7 pounds per square inch absolute, 100 percent oxygen atmosphere.

DETERMINATION:
    The test conditions were extremely hazardous.

RECOMMENDATION:
    The amount and location of combustible materials in the Command Module must be severely restricted and controlled.

3. FINDING:
    a. The rapid spread of fire caused an increase in pressure and temperature which resulted in rupture of the Command Module and creation of a toxic atmosphere. Death of the crew was from asphyxia due to inhalation of toxic gases due to fire. A contributory cause of death was thermal burns.
    b. Non-uniform distribution of carboxyhemoglobin was found by autopsy.

DETERMINATION:
    Autopsy data leads to the medical opinion that unconsciousness occurred rapidly and that death followed soon thereafter.

4. FINDING:
    Due to internal pressure, the Command Module inner hatch could not be opened prior to rupture of the Command Module.

DETERMINATION:
    The crew was never capable of effecting emergency egress because of the pressurization before rupture and their loss of consciousness soon after rupture.

RECOMMENDATION:
    The time required for egress of the crew be reduced and the operations necessary for egress be simplified.

5. FINDING:
    Those organizations responsible for the planning, conduct and safety of this test failed to identify it as being hazardous. Contingency preparations to permit escape or

rescue of the crew from an internal Command Module fire were not made.

a. No procedures for this type of emergency had been established either for the crew or for the spacecraft pad work team.

b. The emergency equipment located in the White Room and on the spacecraft work levels was not designed for the smoke condition resulting from a fire of this nature.

c. Emergency fire, rescue and medical teams were not in attendance.

d. Both the spacecraft work levels and the umbilical tower access arm contain features such as steps, sliding doors and sharp turns in the egress paths which hinder emergency operations.

DETERMINATION:

Adequate safety precautions were neither established nor observed for this test.

RECOMMENDATIONS:

a. Management continually monitor the safety of all test operations and assure the adequacy of emergency procedures.

b. All emergency equipment (breathing apparatus, protective clothing, deluge systems, access arm, etc.) be reviewed for adequacy.

c. Personnel training and practice for emergency procedures be given on a regular basis and reviewed prior to the conduct of a hazardous operation.

d. Service structures and umbilical towers be modified to facilitate emergency operations.

6. FINDING:

Frequent interruptions and failures had been experienced in the overall communication system during the operations preceding the accident.

DETERMINATION:

The overall communication system was unsatisfactory.

RECOMMENDATIONS:

a. The Ground Communication System be improved to assure reliable communications between all test elements as soon as possible and before the next manned flight.

b. A detailed design review be conducted on the entire spacecraft communication system.

***

9. FINDING:
    The Command Module Environmental Control System design provides a pure oxygen atmosphere.
    DETERMINATION:
    This atmosphere presents severe fire hazards if the amount and location of combustibles in the Command Module are not restricted and controlled.
    RECOMMENDATIONS:
    a. The fire safety of the reconfigured Command Module be established by full-scale mock-up tests.
    b. Studies of the use of a diluent gas be continued with particular reference to assessing the problems of gas detection and control and the risk of additional operations that would be required in the use of a two gas atmosphere.

10. FINDING:
    Deficiencies existed in Command Module design, workmanship and quality control, such as:
    a. Components of the Environmental Control System installed in Command Module 012 had a history of many removals and of technical difficulties including regulator failures, line failures and Environmental Control Unit failures. The design and installation features of the Environmental Control Unit makes removal or repair difficult.
    b. Coolant leakage at solder joints has been a chronic problem.
    c. The coolant is both corrosive and combustible.
    d. Deficiencies in design, manufacture, installation, rework and quality control existed in the electrical wiring.
    e. No vibration test was made of a complete flight-configured spacecraft.
    f. Spacecraft design and operating procedures currently require the disconnecting of electrical connections while powered.
    g. No design features for fire protection were incorporated.

DETERMINATION:

These deficiencies created an unnecessarily hazardous condition and their continuation would imperil any future Apollo operations.

RECOMMENDATIONS:

a. An in-depth review of all elements, components and assemblies of the Environmental Control System be conducted to assure its functional and structural integrity and to minimize its contribution to fire risk.

b. Present design of soldered joints in plumbing be modified to increase integrity or the joints be replaced with a more structurally reliable configuration.

c. Deleterious effects of coolant leakage and spillage be eliminated.

d. Review of specifications be conducted, 3-dimensional jigs be used in manufacture of wire bundles and rigid inspection at all stages of wiring design, manufacture and installation be enforced.

e. Vibration tests be conducted of a flight-configured spacecraft.

f. The necessity for electrical connections or disconnections with power on within the crew compartment be eliminated.

g. Investigation be made of the most effective means of controlling and extinguishing a spacecraft fire. Auxiliary breathing oxygen and crew protection from smoke and toxic fumes be provided.

11. FINDING:

An examination of operating practices showed the following examples of problem areas:

a. The number of the open items at the time of shipment of the Command Module 012 was not known. There were 113 significant Engineering Orders not accomplished at the time Command Module 012 was delivered to NASA; 623 Engineering Orders were released subsequent to delivery. Of these, 22 were recent releases which were not recorded in configuration records at the time of the accident.

b. Established requirements were not followed with regard to the pre-test constraints list. The list was not

completed and signed by designated contractor and
NASA personnel prior to the test, even though oral
agreement to proceed was reached.

c. Formulation of and changes to pre-launch test requirements for the Apollo spacecraft program were
unresponsive to changing conditions.

d. Non-certified equipment items were installed in the
Command Module at time of test.

e. Discrepancies existed between NAA and NASA MSC
specifications regarding inclusion and positioning of
flammable materials.

f. The test specification was released in August 1966
and was not updated to include accumulated changes
from release date to date of the test.

DETERMINATION:

Problems of program management and relationships between Centers and with the contractor have led in some
cases to insufficient response to changing program requirements.

RECOMMENDATION:

Every effort must be made to insure the maximum clar-
ification and understanding of the responsibilities of all the
organizations involved, the objective being a fully coordinated and efficient program.

*Eventually the furor over the accident quieted. There were no se-
rious suggestions that the Apollo program be halted or the "before the decade is out" goal be abandoned. At the Manned
Spacecraft Center, George Low assumed management control of
the Apollo spacecraft after the Apollo 1 fire. Under his close supervision, North American set about remedying the deficiencies
in the Apollo spacecraft. Grumman was moving ahead with its
work on the lunar module but continuing to confront both
schedule and weight problems. The Saturn V had its first test
launch on November 9, 1967, on what was designated as the
Apollo 4 mission; all test objectives were met successfully. Apollo
5 and Apollo 6 were additional tests without a crew aboard.
By the beginning of 1968, NASA was ready to schedule the
first launch of the redesigned Apollo command and service module; the date was finally set for October 7. The Earth-orbiting*

Apollo 7 mission would be the first in a sequence of crewed missions leading up to a lunar landing. The missions were designated in NASA internal planning by letters of the alphabet:

C—test of the Apollo command and service module in low Earth orbit;

D—test of the Apollo command and service and lunar modules in low Earth orbit;

E—test of the Apollo command and service and lunar modules in a mission beyond Earth orbit, but not headed to the Moon;

F—test of all equipment in lunar orbit;

G—lunar landing mission.

It was not clear as 1968 began whether following this schedule would provide adequate assurance that the United States would reach the Moon before the Soviet Union. Throughout the 1960s, the Central Intelligence Agency had closely monitored the progress of the Soviet space program. In 1963 U.S. photoreconnaissance satellites had detected what appeared to be the beginning of a large construction project at the main Soviet launch site, the Baikonur Cosmodrome in the Soviet republic of Kazakhstan. By 1964 construction of a large assembly building and two launchpads could be seen. It was during that year that the Soviet leadership finally approved a Soviet Moon program, but there were continuing bureaucratic battles inside the Soviet space community that slowed progress and resulted in inadequate funding for the project. By mid-1965 the U.S. intelligence community had concluded that the Soviet Union did indeed have a lunar program, but that it was not proceeding on a pace that was competitive with Apollo. In December 1967 a U.S. satellite returned an image of a previously unseen large booster on one of the new launchpads.

The reality was that by 1967 the Soviet Union was conducting two Moon programs, one aimed at a lunar landing and a second, using a version of the proven Proton launch vehicle and a modified Soyuz spacecraft called Zond, aimed at flights around the Moon, without the capability to land or even to go into lunar orbit. In April 1968 the Central Intelligence Agency issued an update of a 1967 assessment of the Soviet program. According to the CIA, the United States was well in the lead in achieving the *first lunar landing. Of particular note, however, is the estimate that the Soviet Union might attempt a circumlunar flight with cosmonauts aboard before the end of 1968 or in early 1969. Senior NASA officials were certainly aware of this possibility as they considered in mid-1968 whether to approve sending the Apollo 8 mission into orbit around the moon in December 1968. This intelligence assessment, originally classified "Top Secret," was declassified only in 1997.*

Director of Central Intelligence, "The Soviet Space Program," April 4, 1968

THE PROBLEM

To examine significant developments in the Soviet space program since the publication of NIE 11-1-67, "The Soviet Space Program," dated 2 March 1967, TOP SECRET, and to assess the impact of those developments on future Soviet space efforts with particular emphasis on the manned lunar landing program.

DISCUSSION

1. In the year since publication of NIE 11-1-67, the Soviets have conducted more space launches than in any comparable period since the program began. Scientific and applied satellites, particularly those having military applications, largely account for the increased activity. The Soviets also intensified efforts to develop what we believe to be a fractional orbit bombardment system (FOBS). The photoreconnaissance program continued at the same high rates of the previous two years.

2. In general, the Soviet space program progressed along the lines of our estimate. It included the following significant developments: new spacecraft and launch vehicle development, rendezvous and docking of two unmanned spacecraft, an unsuccessful manned flight attempt (which ended in the death of Cosmonaut Komarov), the successful probe to Venus, an unmanned circumlunar attempt which failed, and a simulated circumlunar mission. Evidence of the past year indicates that the Soviets are continuing to work

toward more advanced missions, including a manned lunar landing, and it provides a better basis for estimating the sequence and timing of major events in the Soviet space program.

3. Considering additional evidence and further analysis, we continue to estimate that the Soviet manned lunar landing program is not intended to be competitive with the US Apollo program. We now estimate that the Soviets will attempt a manned lunar landing in the latter half of 1971 or in 1972, and we believe that 1972 is the more likely date. The earliest possible date, involving a high risk, failure-free program, would be late in 1970. In NIE 11-1-67 we estimated that they would probably make such an attempt in the 1970–1971 period; the second half of 1969 was considered the earliest possible time.

4. The Soviets will probably attempt a manned circumlunar flight both as a preliminary to a manned lunar landing and as an attempt to lessen the psychological impact of the Apollo program. In NIE 11-1-67, we estimated that the Soviets would attempt such a mission in the first half of 1968 or the first half of 1969 (or even as early as late 1967 for an anniversary spectacular). The failure of the unmanned circumlunar test in November 1967 leads us now to estimate that a manned attempt is unlikely before the last half of 1968, with 1969 being more likely. The Soviets soon will probably attempt another unmanned circumlunar flight.

5. Within the next few years the Soviets will probably attempt to orbit a space station which could weigh as much as 50,000 pounds, could carry a crew of 6–8 and could remain in orbit for a year or more. With the Proton booster and suitable upper staging they could do so in the last half of 1969, although 1970 seems more likely. Alternatively, the Soviets could construct a small space station by joining several spacecraft somewhat earlier—in the second half of 1968 or 1969—to perform essentially the same functions. We previously estimated that the earliest the Soviets could orbit such a space station was late 1967 with 1968 being more likely.

6. We continue to believe that the Soviets will establish a large, very long duration space station which would probably

weigh several hundred thousand pounds and would be capable of carrying a crew of 20 or more. Our previous estimate, which gave 1970–1971 as the probable date and late 1969 as the earliest possible, was based primarily upon launch vehicle capacity. We now believe that the pacing item will be the highly advanced life support/environmental control technology required, and that such a station will probably not be placed in orbit before the mid-1970's.

While by 1968 the redesigned Apollo spacecraft seemed ready for a crewed launch, the same could not be said of the Saturn V or the lunar module. The second test launch of the Saturn V, designated Apollo 6, took place on April 4, 1968. In contrast to the almost perfect first test launch the preceding November, there were multiple problems with this flight. Each of the three stages of the vehicle had a separate failure. It took all the skill and experience of the von Braun rocket team to diagnose the causes of the failures. This was essential, because NASA's planning called for the next flight of the Saturn V to carry three astronauts; that would be the "D" mission in NASA's plans.

However, the lunar module scheduled to be flown on that mission arrived at the Kennedy Space Center with a number of problems, and as NASA attempted to address them it appeared increasingly unlikely that the lunar module would be ready to fly in 1968, and indeed that the test flight might not be possible until February or March 1969. If that happened, the likelihood of landing on the Moon by the end of 1969 became remote. Faced with this situation, George Low began to consider an alternative flight sequence: "the possibility of a circumlunar or lunar orbit mission during 1968," using only the command and service module launched by a Saturn V, "as a contingency mission to take a major step forward in the Apollo Program."

By the morning of August 9, as problems with the lunar module persisted, Low took this idea to the director of the Manned Spacecraft Center, Robert Gilruth, who immediately saw its benefits. By the time the day was over, the conversation involved key Apollo decision makers in Houston, Huntsville, and Washington. Although final approval of the preliminary decisions made on August 9 would be several months in coming, it is remarkable that the basics of such a momentous choice could be put in place in just a few hours on one day, and then put in motion a few days later.

George M. Low, "Special Notes for August 9, 1968, and Subsequent," August 19, 1968

BACKGROUND:

<u>June, July 1968.</u> The current situation in Apollo was that LM-3 had been delivered to KSC somewhat later than anticipated; and CSM [command service module] 103 would be delivered to KSC [Kennedy Space Center] in late July. Checkout of 101 at KSC was proceeding well, and a launch in the Fall of 1968 appeared to be assured. There was every reason to believe that 103 would also be a mature spacecraft but that for many reasons LM-3 [lunar module] might run into difficulties. Certification tests of LM were lagging; there were many open failures; and the number of changes and test failures at KSC was quite large.

It had been clear for some time that a lunar landing in this decade could be assured only if the AS 503/CSM 103/LM-3 mission could be flown before the end of 1968. During the June–July time period the projected launch date had slipped from November into December, and the December date was by no means assured. The over-all problem was compounded by the Pogo anomaly [up and down vibrations in the Saturn V launch vehicle] resulting from the Apollo 6 mission, and this remained a significant unknown.

In this time period also the possibility of a circumlunar or lunar orbit mission during 1968, using AS 503 and CSM 103, first occurred to me as a contingency mission to take a major step forward in the Apollo Program.

***

<u>August 7, 1968.</u> With the background of open work and continued problems on LM-3 and the real concern that the mission might not be able to fly until February or March, 1969, I asked Chris Kraft to look into the feasibility of a lunar orbit mission on AS 503 with CSM 103 and without a LM.

***

STEPS IN PLANNING THE MISSION:

August 9, 1968. Met with Gilruth at 0845 and reported to him the detailed status of LM-3 and CSM 103 and informed him that I had been considering the possibility of an AS 503 lunar orbit mission. Gilruth was most enthusiastic and indicated that this would be a major step forward in the program.

Met with Chris Kraft at 0900, and he indicated that his preliminary studies had shown that the mission was technically feasible from the point of view of ground control and onboard computer software. (A step of major importance to make this possible had been taken several months ago when we had decided to use the Colossus onboard computer program for the 103 spacecraft.)

At 0930, I met with Gilruth, Kraft and Slayton. After considerable discussion, we agreed that this mission should certainly be given serious consideration and that we saw no reason at the present time why it should not be done. We immediately decided that it was important to get both von Braun and Phillips on board in order to obtain their endorsement and enthusiastic support. Gilruth called von Braun, gave him the briefest description of our considerations, and asked whether we could meet with him in Huntsville that afternoon. I called Phillips at KSC and also informed him of our activities and asked whether he and Debus could join us in Huntsville that afternoon. Both von Braun and Phillips indicated their agreement in meeting with us, and we set up a session in Huntsville for 2:30 p.m.

August 9, 1968. 2:30 p.m. Met in von Braun's office with von Braun, Rees, James and Richard from MSFC; Phillips and Hage from OMSF; Debus and Petrone from KSC; and Gilruth, Kraft, Slayton and Low from MSC. I described the background of the situation, indicating that LM-3 has seen serious delays and that presently we were one week down on the KSC schedule, indicating a 31 December launch. I went on to indicate that, under the best of circumstances, given a mature spacecraft, we might expect a launch at the end of January; however, with the present situation on LM, I would expect that the earliest possible D mission launch date would be during the middle of March. It therefore appeared that getting all of the benefits of the F (lunar orbit) mission before the D mission was both technically and programmatically advisable. Under this concept a lunar orbit mission, using AS 503 and CSM 103, could be flown in December, 1968. The most significant milestone in this plan would have to be an extremely successful C mission, using CSM 101. However, if 101 were not completely successful, an alternate to the proposed mission would be a CSM alone flight, still in December, using AS 503 and CSM 103 in an earth orbital flight rather than a lunar orbit flight. Under this plan the D mission would be flown on AS 504 with CSM 104 and LM-3, probably still in mid-March. In other words, we would get an extra mission in ahead of the D mission; would get the earliest possible Pogo flight; and would get much of the information needed from the F mission much earlier than we could otherwise. Chris Kraft made the strong point that, in order to gain the F mission flight benefits, the flight would have to be into lunar orbit as opposed to circumlunar flight.

During the remainder of the meeting in Huntsville, all present exhibited a great deal of interest and enthusiasm for this flight.

Phillips outlined on the blackboard the actions that would have to be taken over the next several days. Generally, KSC indicated that they could support such a mission by December 1, 1968; MSFC could see no difficulties from their end; MSC's main concern involved possible differences between CSM 103 and CSM 106, which was the first one that had been scheduled to leave earth orbit, and finding a substitute for the LM for this flight.

The Huntsville meeting ended at 5 p.m. with an agreement to get together in Washington on August 14, 1968. At that time the assembled group planned to make a decision as to whether to proceed with these plans or not. If the decision was affirmative, Phillips would immediately leave for Vienna to discuss the plans with Mueller and Webb, since it would be most important to move out as quickly as possible once the plan was adopted. It was also agreed to classify the planning stage of this activity secret, but it was proposed that, as soon as the Agency had adopted the plan, it should be fully disclosed to the public.

**

<u>August 13, 1968</u> . . . Slayton had decided to assign the 104 crew to this mission (Borman, Lovell and Anders, backed up by

Armstrong, Aldrin and Haise) in order to minimize possible effects on the D mission. Slayton had talked to Borman on Saturday and found him to be very much interested in making this flight.

August 14, 1968. Went to Washington with Gilruth, Kraft and Slayton to meet with Paine, Phillips, Hage, Schneider and Bowman from Headquarters; von Braun, James and Richard from Marshall; and Debus and Petrone from KSC. The meeting started with an MSC review of spacecraft, flight operations, and flight crew support for the proposed mission. I reviewed the Spacecraft 103 hardware configuration, the proposed LM substitute, consumable requirements, and the proposed alternate mission . . .

Kraft indicated that there were no major problems with either the MSFN [Manned Space Flight Network] or the Mission Control Center and flight operations. He discussed the launch window constraints and indicated that NASA management would have to get with the Department of Defense in order to obtain recovery support. Our conclusion was that we should go for the December 20, 1968, launch window with a built-in two week hold prior to the launch . . .

MSFC indicated that there were no significant difficulties with the launch vehicle to support this mission. We agreed that LTA-B would be loaded for a total payload weight of 85,000 pounds. MSFC also agreed that they could provide telemetry for the LTA-B measurements.

Petrone outlined his plans for activities at KSC and indicated that the earliest possible launch date would be December 6, 1968. Other dates included the first manned altitude chamber run on September 14; the move to the VAB on September 28; and move to the pad on October 1.

We also discussed the mission sequence to be followed after the proposed mission and proposed that the best plan would be to fly the D mission next, followed by an F mission, which, in turn, would be followed by the first lunar landing mission. In other words, the proposed mission would take the place of the E mission but would be flown before D. MSC also proposed that for internal planning purposes we should schedule the D mission for March 1, 1969; the F mission for May 15, 1969; and the G mission for July or August, 1969. However, dates two weeks later for D, one month later for F, and one month later for G should be our public commitment dates.

During the course of the meeting Phillips received a call from George Mueller in Vienna. Apparently Phillips had discussed the proposal with Mueller on the previous day, and after thinking it over, Mueller's reception was very cool. Mueller was concerned over stating the plan before the flight of Apollo 7 and was against announcing a plan as we might have to back away from it if 101 did not work. He also indicated that Phillips' arrival and departure in Vienna might create problems with the press and therefore urged Phillips not to come. Mueller's plans were to return to the country on August 21 for a speech in Detroit, and he would not be able to meet with us in Washington until August 22.

All present indicated that we would have to move out immediately in order to meet the December launch window and that a delay until August 22 or later would automatically mean the mission would have to slip until January. It was also hard for us to believe that Mueller was unwilling to accept the plan which was unanimously accepted by all Center Directors and Program Managers. We again urged Phillips to review our findings with Mueller and make a strong plea to visit Mueller in Vienna immediately, assuming, of course, that it was not possible for Mueller to return to this country. We also pointed out that if we were to implement our plan with any degree of confidence, so many people would have to become involved that it would be impossible to keep it quiet for very long.

Following the over-all discussions of the mission, Dr. Paine indicated that it had not been too long since we were uncertain as to whether the Apollo 503 mission should even be manned. Now we were proposing an extremely bold mission. Had we really considered all of the implications? He specifically wanted to know whether anyone present was against making this move. In going around the table, one by one, the following comments were made:

von Braun: Once a decision has been made to fly a man on 503, it doesn't matter to the launch vehicle how far we go. From the program point of view, this mission appears to be simpler than the D mission. The mission should by all means be undertaken.

Hage: There are a number of way stations in the mission. Decision points can be made at each of these way stations, thereby minimizing the over-all risk. I am all for the mission.

Slayton: This is the only chance to get to the moon before the end of 1969. It is a natural thing to do in Apollo today. There are many positive factors and no negative ones.

Debus: I have no technical reservations; however, it will be necessary to educate the public, for if this is done wrong and we fail, Apollo will have a major setback. By all means fly the mission.

Petrone: I have no reservations.

Bowman: It is a shot in the arm for manned space flight.

James: Manned safety in this flight and in the following flights is enhanced. The over-all Apollo budget and schedule position is enhanced. An early go-ahead is needed.

Richard: The decision to fly manned has already been made for 503. Our lunar capability in Apollo is enhanced by flying this mission now.

Schneider: This has my whole-hearted endorsement. There are very valid reasons for pressing on.

Gilruth: Although this may not be the only way to make our goal, it certainly enhances our possibility. There is always risk in manned space flight, but this is a path of less risk. In fact, it has a minimum risk of all of our Apollo plans. If I had the key decision, I would make it in the affirmative.

Kraft: Probably the flight operations people have the most difficult job in this. We will need all kinds of priorities. It will not be easy to do, but I have every confidence in our doing it. However, it should be a lunar orbit mission and not a circumlunar mission.

Low: This is really the only thing to do technically in the current state of Apollo. Assuming a successful Apollo 7 mission, there is no other choice. The question is not whether we can afford to do it, it should be can we afford not to do it.

Following this set of comments, Paine congratulated the assembled group for not being prisoners of previous plans and indicated that he personally felt that this was the right thing for Apollo and that, of course, he would have to work with Mueller and Webb before it could be approved.

Phillips indicated that his conclusion was that this was a technically sound thing to do and does not represent a short cut introducing additional risks. Our plan would be to meet with Mueller on Thursday, August 22, in Washington. Phillips reiterated Mueller's reservations. These included reservations about program risks such as possible questions about irresponsible scheduling, possible program impact if the Apollo 7 mission should fail and we could not proceed with an announced major step forward, and the question concerning program impact of a catastrophic failure on this special mission.

At the conclusion of the meeting we agreed to move out on a limited basis. Since the day-by-day timing was critical, Phillips agreed that we should involve the next level of people required to carry forward with our plans, giving them, of course, proper instructions about the current security classification of the mission. At the conclusion of the meeting Phillips indicated the earliest possible decision would come in 7 to 10 days under the best of circumstances.

***

After much discussion, we finally decided that the most important thing Apollo can achieve this year is a lunar capability in hardware, software, crew training, etc. This, we believe, is necessary whether the C´ mission goes to the moon or not. We also agreed that the only way to achieve this lunar capability is to plan the mission as though it were going to fly to the moon. By so doing, all involved would, without question, have to face the real issues and make the real decisions that would allow us to go to the moon. An earth orbital mission would, of course, be a natural fallout because such a mission would have to be an abort option for a lunar mission in the event that the S-IVB stage could not make its second burn. Therefore, by planning such a mission, we would have, in December, an earth orbital capability on the C´ mission while at the same time having completed all the planning and preparation that would be necessary should conditions be such that we could go to the moon. We would not commit now, either within NASA or outside, to do any more than the earth orbital mission.

This plan was adopted, and the over-all program plan can best be summarized as follows:

a. AS-503, designated Apollo 8, will be prepared to be ready for launch on December 6, 1968. It will consist of CSM 103, LTA-B, and AS-503. The reasons for making the change from the previously defined mission are that this will give us the earliest possible Pogo checkout flight and that LM checkout delays have prevented us from making an early flight with the LM.

b. The mission will be designated as C′. It will be an earth orbital mission, including whatever elements of C need to be repeated and elements of D, E, F, and G that can be incorporated.

c. Final definition of the mission will not come after Apollo 7.

d. The crew will be the E mission crew so that the D mission crew can continue its active preparation for that mission.

e. We recognize that after the C′ mission the remaining missions will be upon us and that it is essential to bring lunar capability into being while we are implementing the C′ mission. This includes lunar capability in hardware, software, flight operations, and crew operations.

f. This capability can only be brought into being if we plan for it now. Therefore, we will do all of our planning for the C′ mission as though it were a lunar orbit mission. This will give us maximum flexibility to fly the assigned earth orbital mission with whatever elements of all other missions, including the lunar landing mission, are best to put into that flight after the results of Apollo 7 are known.

While the Apollo managers could begin to plan for a lunar mission, they could not commit NASA to undertaking such a bold step until the October mission, designated Apollo 7, was a success. The Apollo 7 mission took place from October 11 to October 22. All objectives of the flight were met, clearing the path for a decision to send the Apollo 8 mission into lunar orbit.

That decision would not be made by James E. Webb. On September 16, Webb had gone to the White House for a meeting with President Johnson to discuss a variety of issues, including how best to protect NASA and particularly Apollo during the transition to the next president. (Lyndon Johnson had announced in March 1968 that he would not seek reelection.) Webb was weary after six and a half years running NASA at a frenetic pace, and had been a target of congressional criticism since the Apollo 1 fire. Webb thought that at some point in the fall he should step aside and let Thomas Paine, a nonpolitical person, demonstrate that he was capable of running NASA at least through the first lunar landing. To Webb's surprise, the president not only accepted Webb's offer of resignation, but decided that Webb should announce it immediately, effective on Webb's sixty-second birthday, October 7.

Although momentum was great after the success of Apollo 7, a final decision to undertake to send Apollo 8 into lunar orbit had not yet been made. A final review of the mission was scheduled for November 10 and 11. The November 10 meeting included the top executives of the companies involved in Apollo. After hearing a series of presentations by NASA managers, the executives were polled on their views of whether Apollo 8 should be approved as a lunar orbit mission. Although a few questions were raised, according to George Low, "The meeting was adjourned with the conclusion that a firm recommendation to fly the Apollo 8 mission to lunar orbit would be made the next day to the Acting Administrator." On November 11 there were a series of internal NASA meetings to discuss the mission. At their conclusion, Acting Administrator Paine announced that he had approved the plan to make Apollo 8 a mission to go into orbit around the Moon. The launch date was set for December 21, which meant that the Apollo spacecraft would go into lunar orbit on Christmas Eve.

The five first-stage engines of the Saturn V booster rumbled into action at 7:51 a.m. on December 21, lifting the Apollo 8 crew of Frank Borman, James Lovell, and William Anders on their historic journey. Less than three hours later, the engine on the third stage of the launch vehicle fired, injecting the Apollo 8 spacecraft on a trajectory that would take it to the vicinity of the Moon three days later. For the first time in history, humans had left Earth orbit to venture into deep space. Once the Apollo 8 spacecraft arrived at the Moon, the engine on its service module fired, placing the Apollo spacecraft into lunar orbit, where it would remain for twenty hours.

The public highlight of the mission came on Christmas Eve, as the crew televised their view of the lunar surface back to millions

*of people on Earth. Then, to the surprise of almost everyone, in-
cluding the mission controllers back on Earth, the crew took
turns reading the first verses from the Bible's Genesis account of
the creation of the Earth. Frank Borman closed their broadcast
by saying "Goodnight, good luck, a Merry Christmas, and God
bless you all—all of you on the good Earth." In addition to this
dramatic broadcast, the Apollo 8 crew brought home with them
the iconic "Earthrise" photograph of the blue Earth rising above
the desolate lunar landscape. The successful mission demonstrated that NASA was ready to operate at the lunar distance.*

*Two missions stood between Apollo 8 and, if they were suc-
cessful, the first attempt at a lunar landing. On March 3, 1969, a
Saturn V launched the full Apollo spacecraft for the first time—
the command and service module and the lunar module. Over
the course of the ten-day mission in Earth orbit, the lunar module spent six hours undocked from the command and service
module, traveling up to 113 miles distant before returning to
rendezvous and re-dock, thereby demonstrating an essential element of the lunar orbital rendezvous approach. Both the descent
and ascent engines of the lunar module were fired in a variety of
modes. The mission was extremely complex, and all of its objectives were met successfully.*

*Apollo 10 would be a dress rehearsal for the lunar landing
mission, carrying out all elements of that mission except for the
final descent from 47,000 feet above the lunar surface. Once
again, Apollo 10 met all of its test objectives. The mission was
launched on May 18 and returned to a safe landing in the Pacific
Ocean on May 26. Apollo 11 was next.*

*Chief astronaut Deke Slayton had adopted an approach to
flight crew assignment that resulted in the backup crew for a particular mission becoming the prime crew for a mission three
flights down the line. That meant that the Apollo 11 flight assignment would go to the crew that had been the backup for
Apollo 8—Neil Armstrong, Edwin "Buzz" Aldrin, and Fred
Haise, a replacement for Michael Collins, who had recently had
back surgery. On January 6, 1969, Slayton informed Neil Armstrong that he would command the Apollo 11 mission, with Aldrin as his lunar module pilot. By that time, Collins had fully
recovered from the surgery that had sidelined him as backup for*

Apollo 8, and he, rather than Haise, would serve as command module pilot.

One matter of concern as Apollo 11 approached was what might be said as the first man stepped onto the Moon. Julian Scheer, the top public affairs official at NASA Headquarters, heard a rumor that George Low was seeking advice on what Armstrong might say and wrote him to discuss the matter. In his response to Scheer, Low mentions Si Bourgin, who was an employee of the U.S. Information Agency with a particular focus on the space program; it had been Bourgin's wife who first suggested that the Apollo 8 crew read from the Bible as they orbited the Moon on Christmas Eve 1968.

Letter to George M. Low (Manager, Apollo Spacecraft Program), from Julian Scheer (Assistant Administrator for Public Affairs), March 12, 1969

Dear George:

It has come to my attention that you have asked someone outside of NASA to advise you on what the manned lunar landing astronauts might say when they touch down on the Moon's surface. This disturbs me for several reasons.

The Agency has solicited from within NASA any suggestions on what materials and artifacts might be carried to the surface of the Moon on that historic first flight. But we have not solicited comment or suggestions on what the astronauts might say. Not only do I personally feel that we ought not to coach the astronauts, but I feel it would be damaging for the word to get out that we were soliciting comment. The ultimate decision on what the astronauts will carry is vested in a committee set up by the Administrator; the committee will not, nor will the Agency by any other means, suggest remarks by the astronauts.

Frank Borman solicited a suggestion from me on what would be appropriate for Christmas Eve. I felt—and my feeling still stands—that his reading from the Bible would be diminished in the eyes of the public if it were thought that NASA pre-planned such a thing. I declined both officially and personally to suggest words to him despite the fact that I had some ideas. I believed then and I believe now the same is true of the Apollo 11 crew—that the truest emotion at the historic moment is what the explorer feels within himself, not for the astronauts to be coached before they leave or to carry a prepared text in their hip pocket.

The Lunar Artifacts Committee, chaired by Willis Shapley, asked that all elements of NASA consider what might be carried on Apollo 11. I know that General Phillips has properly reiterated the request by asking all elements of Manned Flight to suggest things, but it was not the desire or intent of the committee to broaden the scope of the solicitation to verbal reactions.

There may be some who are concerned that some dramatic utterance may not be emitted by the first astronaut who touches the lunar surface. I don't share that concern. Others believe a poet ought to go to the Moon. Columbus wasn't a poet and he didn't have a prepared text, but his words were pretty dramatic to me. When he saw the Canary Islands [what is now called the Bahama Islands] he wrote, "I landed, and saw people running around naked, some very green trees, much water, and many fruits."

Two hundred years before Apollo 8, Captain James Cook recorded while watching the transit of Venus over the sun's disk, "We very distinctly saw an atmosphere or dusky shade around the body of the planet."

Meriwether Lewis, traveling with William Clark, recorded, "Great joy in camp. We are in view of the ocean, this great Pacific Ocean which we have been so long anxious to see, and the roreing or noise made by the waves brakeing on the rockey shore may be heard distinctly."

Peary was simply too tired to say anything in 1909 when he reached the North Pole. He went to sleep. The next day he recorded in a diary, "The pole at last. The prize of three centuries. I cannot bring myself to realize it. It seems all so simple and commonplace."

The words of these great explorers tell us something of the men who explore and it is my hope that Neil Armstrong or Buzz Aldrin will tell us what they see and think and nothing that we feel they should say.

I have often been asked if NASA indeed plans to suggest comments to the astronauts. My answer on behalf of NASA is "no."

Letter to Julian Scheer (Assistant Administrator for Public Affairs), from George M. Low (Manager, Apollo Spacecraft Program), March 18, 1969

Dear Julian:

I have just received your letter of March 12, 1969, which apparently stemmed from a misunderstanding. Let me first point out that I completely agree with you that the words said by the astronauts on the lunar surface (or, for that matter, at any other time) must be their own. I have always felt that way and continue to do so.

I am, of course, aware of the Shapley Committee that was established by Dr. Paine, and have also received a copy of a telegram from General Phillips soliciting our comments on what should be carried to the lunar surface. I felt that in order to respond properly to General Phillips and to the Shapley Committee, I would like to seek the advice of Si Bourgin, whose judgment I respect a great deal in these matters. As you know, I met Si on our trip to South America and found that he offered excellent advice to all of us throughout our trip. I, therefore, called Si as soon as he returned from Europe and asked him whether he could offer any advice concerning what the astronauts should *do* (not *say*) when we have first landed on the moon. Si called me back the night before the Apollo 9 launch, and we discussed his ideas at some length. We again agreed at that time that it is properly NASA's function to plan what artifacts should be left on the lunar surface or what should be brought back, but that the words that the astronauts should say must be entirely their own.

Since then, I have had a meeting with Neil Armstrong to discuss with him some of our ideas and suggestions, including those of Si Bourgin's, in order to solicit his views. Even though I had not yet received your letter at that time, we also discussed the point that whatever things are left on the lunar surface are things that he must be comfortable with, and whatever words are said must be his own words.

All of these activities—my discussions with Si, my discussions with Neil, and discussions with many others within and outside of NASA—are to gain the best possible advice that I can seek for what I consider to be a most important event. The result of all of this will be my input to Dr. Gilruth so that he can forward it to the Shapley Committee, should he so desire.

I hope that this clarifies any misunderstanding that we might have had on this matter.

Once the Apollo 11 prime crew had been chosen, there followed almost seven months of intensive training to get them ready for the mission. While they and their colleagues at the Manned Spacecraft Center focused on that training, NASA Headquarters considered how best to attend to the symbolic aspects of the mission.

Richard Nixon was elected president in November 1968, and as he took office he asked Thomas Paine to continue to serve as acting NASA administrator; only after a number of others had turned the job down did Nixon on March 5, 1969, nominate Paine to be administrator on a permanent basis.

During the spring of 1969 Paine appointed one of his top advisers, associate deputy administrator Willis Shapley, to chair a Symbolic Activities Committee to recommend to him how best to recognize the historic character of the first lunar landing. Shapley was a veteran Washington bureaucrat, and Paine looked to him for advice on political, policy, and budgetary issues. By mid-April, the committee made initial decisions regarding what items would be carried to the moon, and what symbolic activities would be carried out on the lunar surface. The final decisions on these matters were communicated to the Apollo program management just two weeks before the July 16 liftoff of the mission. Some of the items would be stowed in the command module (CM), which would remain in lunar orbit as the lunar module (LM) descended to the lunar surface.

Memorandum to Dr. George Mueller from Willis H. Shapley (NASA Associate Deputy Administrator), "Symbolic Items for the First Lunar Landing," April 19, 1969

This is to advise you, the Apollo Program Office, and MSC of the thinking that has emerged from discussions among members of the Symbolic Activities Committee to date on symbolic activities in connection with the first lunar landing, including articles to be left on the moon and articles to be taken to the moon and returned.

Further discussions will be necessary prior to the time we will make final recommendations for decision by the Administrator, and comments and suggestions from all members of the Committee and others are still in order. However, in view of the general agreement on approach that has been manifested so far and the tight deadlines for decisions on matters directly affecting preparations for the mission, the approach outlined below should be taken as the basis for further planning at this time.

1. Symbolic activities must not, of course, jeopardize crew safety or unduly interfere with or degrade achievement of mission objectives. They should be simple, in good taste from a world-wide standpoint, and have no commercial implications or overtones.

2. The intended overall impression of the symbolic activities and of the manner in which they are presented to the world should be to signalize the first lunar landing as an historic forward step of all mankind that has been accomplished by the United States of America.

3. The "forward step of all mankind" aspect of the landing should be symbolized primarily by a suitable inscription to be left on the moon and by statements made on earth, and also perhaps by leaving on the moon miniature flags of all nations. The UN flag, flags of all other regional or international organizations, or other international or religious symbolism will not be used.

4. The "accomplishment by the United States" aspect of the landing should be symbolized primarily by placing and leaving a U.S. flag on the moon in such a way as to make it clear that the flag symbolized the fact that an effort by American people reached the moon, not that the U.S. is "taking possession" of the moon. The latter connotation is contrary to our national intent and would be inconsistent with the Treaty on Peaceful Uses of Outer Space.

5. In implementing the approach outlined above, the following primary symbolic articles and actions or their equivalents should be considered for inclusion in the mission:

   a. A U.S. flag to be placed and left on the moon. The flag should be such that it can be clearly photographed and televised. If possible, the act of emplacing the flag by the astronaut, as well as the emplaced flag with an astronaut beside it, should be photographed and televised. Current thinking is that a recognizable traditional flag should be emplaced on the moon. The flag decal on the LM descent stage would not by itself suffice unless a flag proved to be clearly not feasible.

***

6.   The LM descent stage itself will be of prime symbolic significance since the descent stage will become a permanent monument on the surface of the moon. For this reason, the name given to the LM and any inscriptions to be placed on it must be consistent with the overall approach on symbolic articles and must be approved by the Administrator. The present thinking is that:

   a.   The name of the vehicle should be dignified and hopefully convey the sense of "beginning" rather than "culmination" of man's exploration of other worlds.

***

Memorandum to Dr. George Mueller from Willis Shapley, (NASA Associate Deputy Administrator), "Symbolic Activities for Apollo 11," July 2, 1969

As your office has previously been advised, the symbolic articles approved for the Apollo 11 mission as of this date are as follows:

A.   Symbolic articles to be left on the moon

   1.   A <u>U.S. flag</u>, on a metal staff with an unfurling device, to be emplaced in the lunar soil by the astronauts. This will be the only flag emplanted or otherwise placed on the surface of the moon.

   2.   A <u>commemorative plaque</u> affixed to the LM descent stage to be unveiled by the astronauts. The plaque will be inscribed with:

      a.   A design showing the two hemispheres of the earth and the outlines of the continents, without national boundaries.

      b.   The words: "Here men from the planet earth first set foot upon the moon. We came in peace for all mankind."

      c.   The date (month and year).

      d. The signatures of the three astronauts and the President of the U.S.

   3. A <u>microminiaturized photoprint</u> of letters of good will received from Chiefs of State or other representatives of foreign nations.

B. Symbolic articles to be taken to the moon and returned to earth

   1. <u>Miniature flags</u> (1 each) of all nations of the UN, and of the 50 states, District of Columbia, and U.S. territories—for subsequent presentation as determined by the President. "All nations" has been defined on the advice of the State Department to include "the members of the United Nations and the UN Specialized Agencies." These items will be stowed in the LM.

   2. <u>Small U.S. flags</u>—for special presentation as determined by the President or the Administrator of NASA. These will also be stowed in the LM.

   3. <u>Stamp die</u> from which Post Office Department will print special postage stamps commemorating the first lunar landing and a <u>stamped envelope</u> to be cancelled with the <u>cancellation stamping device</u>. Cancellation can be done as convenient during the mission in the CM. The stamp die will be stowed in the LM; the stamping device and envelope will be stowed in the CM. <u>These items will not be announced in advance</u>.

   4. Two <u>full size U.S. flags</u>—which have been flown over the Capitol, the House and the Senate, to be carried in CM but will not be transferred to the LM.

C. Personal Articles

Personal articles of the astronauts' choosing under arrangements between Mr. Slayton and the flight crews.

With respect to all items under categories A and B above, it should be clearly understood that the articles are "owned" by the Government and that the disposition of the articles themselves or facsimiles thereof is to be determined by the Administrator or NASA. The articles returned from the mission should be turned over to a proper authority at MSC promptly upon return. In the case of Item B2, the Administrator has determined that a reasonable number of small U.S. flags will be made available to the flight crew for presentation as they see fit, subject to the avoidance of conflict with plans for presentation of these flags by the President or the Administrator.

With respect to articles in Category C above, Mr. Scheer should be notified in advance of the mission of any items which are or may appear to be duplicates of items the President or others might present to Governors, Heads of State, etc. The value of these "one-of-a-kind" presentations can be diminished if there is a proliferation of such items. Flags and patches particularly fall into this category.

Public announcement has or will be made of all items in Categories A and B in advance of the mission except for the items under B3, any release concerning which is subject to a separate decision.

The Nixon White House was also preparing for what all recognized would be a historic mission. One suggestion considered and quickly rejected was to recognize that the lunar landing goal had originated with President Kennedy by naming the Apollo 11 spacecraft "John F. Kennedy." The suggestion came from Bill Moyers, a former Kennedy and Johnson aide who in 1969 was editor of the daily newspaper Newsday. *This memorandum from a presidential assistant, Stephen Bull, produced the predictably negative reaction from Nixon's top political assistant, H. R. Haldeman. In fact, Richard Nixon never once publicly mentioned John Kennedy in the course of celebrating the Apollo 11 achievement.*

Memorandum to H. R. Haldeman from Stephen Bull, June 13, 1969

June 13, 1969

TO: H. R. Haldeman

Pat Moynihan has forwarded a proposal set forth by Bill Moyers and Newsday that the Apollo 11 moon shot be commissioned "The John F. Kennedy" (Tab A).

Drs. Burns and DuBridge endorse this proposal. Dr. Burns noted that "Such an act of graciousness is justified by history and would be...good politics...." (Tab B).

Bryce Harlow, John Ehrlichman and Herb Klein replied rather vehemently in opposition to such a proposal. Bryce notes that the nation's entire rocketry program was initiated by President Eisenhower and that we have gone far enough in "Kennedyizing" such ventures. John Ehrlichman notes rather practically that such an action would win us neither friends in Congress nor votes in 1972. John concluded his opinion by stating "Fall prey to this and the next step will be renaming the moon because NBC thinks it would be a good idea" (Tab C).

Stephen Bull

HRH Action:

That any plan to commission the
Apollo 11 shot John F. Kennedy
be abandoned:_____ ~~X.~~ _positively !!_

That we commission the Apollo 11
shot John F. Kennedy:_____

Other _____

To help him prepare for his involvement with Apollo 11, Nixon asked NASA to loan Apollo 8 commander Frank Borman to the White House as his personal assistant with respect to the Apollo 11 mission. Nixon decided not to attend the Apollo 11 liftoff; instead, he would greet the crew upon their return to Earth. In preparation for that meeting and subsequent interactions, Nixon asked Borman to prepare for the president and his wife, Pat Nixon, brief portrayals of the Apollo 11 crew and their wives.

Frank Borman, Memos for the President and the First Lady with respect to the Apollo 11 Crew Members and Their Wives, July 14, 1969

The following is the background information you requested on the Apollo 11 crew:

Commander Neil Armstrong

Born and raised in Ohio. Graduated from Purdue University with Navy scholarship. Was always interested in aviation. Flew 78 combat missions off a carrier in the Korean War; was shot down and rescued. Was test pilot for NASA; flew X-15; flew on board Gemini 8, which was aborted after eight hours because of control malfunction. Wife's name is Jan; has two sons. Quiet, perceptive, thoroughly decent man, whose interests still turn to flying. Has bought interest in both a glider and an airplane, Follows the stock market actively. A little reserved, but when get to know him, he has a very warm personality.

Lunar Module Pilot Edwin ("Buzz") Aldrin

Born and raised in New Jersey. Graduated from West Point. Also flew in Korea—shot down two Migs. Very athletic, aggressive, hard charging. Earned PhD from M.I.T. and was Assistant Dean of the Air Force Academy. Has two sons and a daughter; wife's name is Joan. Almost humorless, a serious personality; very concerned about social problems. Flew very successful mission on Gemini 12, in which he conducted the extra-vehicular activities. His background at M.I.T. contributed greatly to the rendezvous techniques used by NASA.

Command Module Pilot Mike Collins

Comes from a military family. Was born in Rome, Italy; uncle was Lightning Joe Collins; father was also a Major General in the Army; brother was a Brigadier General in the Army. Graduated from West Point. Best handball player among the astronauts. Superb physical condition. Flew for successful Gemini 10, in which he performed the extra-vehicular activities. Was originally scheduled for Apollo 8, but a ruptured disc in his neck caused an operation which result in his removal from that crew. In some senses skeptical. More inclined toward the arts and literature rather than engineering. A devoted family man and also an avid follower of the stock market.

***

The following is some background information on the wives of the Apollo 11 crew:

Jan Armstrong

Wife of Commander Neil Armstrong. Daughter of a doctor. Met Neil at Purdue, where both were students. Has two young sons. Very active in water ballet instruction for neighborhood girls. Quite composed and very factual.

Joan Aldrin

Wife of Lunar Module Pilot, "Buzz" Aldrin. Raised in New Jersey. Mother of two young sons and a daughter. Quite active in drama. Graduated from Douglass College, New Jersey, with a degree in dramatic literature. More demonstrative than either of the other wives, and perhaps more apt to show her concern.

Pat Collins

Wife of Mike Collins. Born and raised in Massachusetts. Mother of two young daughters and a son. Quite active in social work. Met her husband while serving as a recreation director for the Air Force in France. Tends toward the intellectual; very interested in literature and current events. Enjoys evenings that include candlelight and wine for dinner.

The success of Apollo 11 was far from guaranteed, and preparation needed to be made in the event of failure. Nixon speechwriter William Safire prepared these remarks for the president to use in the event that Neil Armstrong and Buzz Aldrin were stranded on the Moon.

William Safire, "In the Event of a Moon Disaster," July 18, 1969

Fate has ordained that the men who went to the moon to explore in peace will stay on the moon to rest in peace.

These brave men, Neil Armstrong and Edwin Aldrin, know

that there is no hope for their recovery. But they also know that there is hope for mankind in their sacrifice.

These two men are laying down their lives in mankind's most noble goal: the search for truth and understanding.

They will be mourned by their families and friends; they will be mourned by their nation; they will be mourned by the people of the world; they will be mourned by a Mother Earth that dared send two of her sons into the unknown.

In their exploration, they stirred the people of the world to feel as one; in their sacrifice, they bind more tightly the brotherhood of man.

In ancient days, men looked at stars and saw their heroes in the constellations. In modern times, we do much the same, but our heroes are epic men of flesh and blood.

Others will follow, and surely find their way home. Man's search will not be denied. But these men were the first, and they will remain the foremost in our hearts.

For every human being who looks up at the moon in the nights to come will know that there is some corner of another world that is forever mankind.

The first mission to land humans on the Moon was launched at 9:32 a.m. EDT on July 16, 1969. Four days later, at 4:17 p.m., the lunar module, after a perilous descent during which Armstrong took over the controls from the module's computer, came to rest on the lunar surface. A few seconds later, Armstrong radioed back, "Houston, Tranquility Base here. The Eagle has landed."

The mission plan called for Armstrong and Aldrin to go to sleep between the time they landed and the time they exited the lunar module for the first Moon walk. But with the landing safely behind them and the lunar module in good condition, the keyed-up crew suggested that they begin their Moon walk five hours ahead of schedule, without the intervening sleep period. Permission was quickly granted. Getting ready to leave the lunar module went more slowly than had been planned, but finally, at 10:56 p.m. EDT, Neil Armstrong stepped off the lunar module, saying, "That's one small step for a man, one giant leap for mankind." Aldrin followed Armstrong fourteen minutes later.

The two spent two and a half hours carrying out their assigned tasks, including planting the U.S. flag on the lunar surface.

During their Moon walk, President Nixon called from the White House.

Richard M. Nixon, Telephone Call to Apollo 11 Astronauts on the Moon, July 20, 1969

Hello Neil and Buzz, I am talking to you by telephone from the Oval Room at the White House, and this certainly has to be the most historic telephone call ever made from the White House.

I just can't tell you how proud we all are of what you have done. For every American this has to be the proudest day of our lives, and for people all over the world I am sure that they, too, join with Americans in recognizing what an immense feat this is.

Because of what you have done the heavens have become a part of man's world, and as you talk to us from the Sea of Tranquility, it inspires us to redouble our efforts to bring peace and tranquility to earth.

For one priceless moment in the whole history of man all the people on this earth are truly one—one in their pride in what you have done and one in our prayers that you will return safely to earth.

* * *

On July 21 Armstrong and Aldrin launched from the lunar surface to rendezvous with Michael Collins, who had been circling the Moon in the command and service module Columbia. *Later in the day the service module engine fired, sending them on a trajectory for a landing in the Pacific Ocean at 12:50 p.m. EDT on July 24. The crew, the command module, and the forty-four pounds of precious lunar cargo were immediately placed in quarantine, where they were soon greeted by the president, who had flown to the recovery ship, the aircraft carrier* Hornet, *to greet them. Nixon, caught up in the excitement of the moment, suggested that the eight days of the Apollo 11 mission were "the greatest week in the history of the world since the Creation." The* Hornet *docked in Honolulu, Hawaii, on the afternoon of July 26; from there, the crew flew back to Houston, but only after clearing U.S. immigration and customs. Like all travelers who return to the United States from trips outside the country, the Apollo 11 crew had to file this declaration as the ship carrying them and their cargo reached their first U.S. port of entry, Honolulu, after their return from the Moon.*

"General Declaration: Agriculture, Customs, Immigration, and Public Health," July 24, 1969

GENERAL DECLARATION

(Outward/Inward)

AGRICULTURE, CUSTOMS, IMMIGRATION, AND PUBLIC HEALTH

Owner or Operator .... NATIONAL AERONAUTICS AND SPACE ADMINISTRATION ....

Marks of Nationality and Registration .... U.S.A. .... Flight No. ... APOLLO 11 .... Date .... JULY 24, 1969 ....

Departure from .... MOON .... Arrival at .... HONOLULU, HAWAII, U.S.A. ....
                        (Place and Country)   (Place and Country)

FLIGHT ROUTING
("Place" Column always to list origin, every en-route stop and destination)

PLACE	TOTAL NUMBER OF CREW	NUMBER OF PASSENGERS ON THIS STAGE	CARGO
CAPE KENNEDY	COMMANDER NEIL A. ARMSTRONG		
MOON	*(signature)*	Departure Place: Embarking ..... NIL Through on same flight ..... NIL	MOON ROCK AND MOON DUST SAMPLES Cargo Manifests Attached
JULY 24, 1969 HONOLULU	COLONEL EDWIN E. ALDRIN, JR. *(signature)*	Arrival Place: Disembarking ..... NIL Through on same flight ..... NIL	
	(signature) LT. COLONEL MICHAEL COLLINS		

Declaration of Health

Persons on board known to be suffering from illness other than airsickness or the effects of accidents, as well as those cases of illness disembarked during the flight:

NONE

Any other condition on board which may lead to the spread of disease:

TO BE DETERMINED

Details of each disinsecting or sanitary treatment (place, date, time, method) during the flight. If no disinsecting has been carried out during the flight give details of most recent disinsecting:

Signed, if required ....
                  Crew Member Concerned

For official use only

HONOLULU AIRPORT
Honolulu, Hawaii
ENTERED

Ernest J. Muraca
Customs Inspector

I declare that all statements and particulars contained in this General Declaration, and in any supplementary forms required to be presented with this General Declaration are complete, exact and true to the best of my knowledge and that all through passengers will continue/have continued on the flight.

Neil Armstrong was a test pilot, and Apollo 11 was in technical terms a test mission. In the days after returning to Houston, he filed a detailed "pilots' report" describing the Apollo 11 mission. This report captures in unemotional prose what actually took place during the historic Apollo 11 lunar landing mission. Included here are portions of the crew's report on mission activities.

Mission Evaluation Team, NASA Manned Spacecraft Center, "Apollo 11: Mission Report"

***

4.0 PILOTS' REPORT

4.1 Prelaunch Activities

All prelaunch systems operations and checks were completed on time and without difficulty. The configuration of the environmental control system included operation of the secondary glycol loop and provided comfortable cockpit temperature conditions.

4.2 Launch

Lift-off occurred precisely on time with ignition accompanied by a low rumbling noise and moderate vibration that increased significantly at the moment of hold-down release. The vibration magnitudes decreased appreciably at the time tower clearance was verified. The yaw, pitch, and roll guidance-program sequences occurred as expected. No unusual sounds or vibrations while passing through the region of maximum dynamic pressure and the angle of attack remained near zero. The S-IC/S-II staging sequence occurred smoothly and at the expected time.

The entire S-II stage flight was remarkably smooth and quiet, and the launch escape tower and boost protective cover were jettisoned normally. The mixture ratio shift was accompanied by a noticeable acceleration decrease. The S-II/S-IVB staging sequence occurred smoothly and approximately at the predicted time. The S-IVB insertion trajectory was completed without incident and the automatic guidance shutdown yielded an insertion-orbit ephemeris, from the command module computer, of 102.1 by 103.9 miles.

Communications between the crewmembers and the Network were excellent throughout all stages of launch.

***

4.10 LUNAR MODULE DESCENT

***

4.10.3 Powered Descent

Ignition for powered descent occurred on time at the minimum thrust level, and the engine was automatically advanced to the fixed throttle point (maximum thrust) after 26 seconds. Visual position checks indicated the spacecraft was 2 or 3 seconds early over a known landmark, but with little cross-range error. A yaw maneuver to a face-up position was initiated at an altitude of about 45,900 feet approximately 4 minutes after ignition. The landing radar began receiving altitude data immediately. The altitude difference, as displayed from the radar and the computer, was approximately 2800 feet.

At 5 minutes 16 seconds after ignition, the first of a series of computer alarms indicated a computer overload condition. These alarms continued intermittently for more than 4 minutes, and although continuation of the trajectory was permissible, monitoring of the computer information display was occasionally precluded . . .

Attitude-thruster firings were heard during each major attitude maneuver and intermittently at other times. Thrust reduction of the descent propulsion system occurred nearly on time (planned at 6 minutes 24 seconds after ignition), contributed to the prediction that the landing would probably be down range of the intended point, inasmuch as the computer had not been corrected for the observed downrange error . . .

After it became clear that an automatic descent would terminate in a boulder field surrounding a large sharp-rimmed crater, manual control was again assumed, and the range was extended to avoid the unsatisfactory landing area. The rate-of-descent mode of throttle (program P66) was entered in the computer to reduce altitude rate so as to maintain sufficient height for landing-site surveillance.

Both the downrange and the crossrange positions were adjusted to permit final descent in a small, relatively level area bounded by a boulder field to the north and sizable craters to the east and south. Surface obscuration caused by blowing dust was apparent at 100 feet and became increasingly severe as the altitude decreased. Although visual determination of horizontal velocity, attitude, and altitude rate were degraded, cues for these variables were adequate for landing. Landing conditions are estimated to have been 1 or 2 ft/sec left, 0 ft/sec forward, and 1 ft/sec down; no evidence of vehicle instability at landing was observed.

***

4.12 LUNAR SURFACE OPERATIONS

***

4.12.2 Egress Preparation

The crew had given considerable thought to the advantage of beginning the extravehicular activity as soon as possible after landing instead of following the flight plan schedule of having the surface operations between two rest periods. The initial rest period was planned to allow flexibility in the event of unexpected difficulty with postlanding activities. These difficulties did not materialize, the crew were not overly tired, and no problem was experienced in adjusting to the 1/6-g environment. Based on these facts, the decision was made at 104:40:00 to proceed with the extravehicular activity prior to the first rest period.

Preparation for extravehicular activity began at 106:11:00. The estimate of the preparation time proved to be optimistic. In simulations, 2 hours had been found to be a reasonable allocation; however, everything had also been laid out in an orderly manner in the cockpit, and only those items involved in the extravehicular activity were present. In fact, items involved in the extravehicular activity were present. In fact, there were checklists, food packets, monoculars, and other miscellaneous items that interfered with an orderly preparation. All these items required some thought as to their possible interference or use in the extravehicular activity. This interference resulted in exceeding the time line estimate by a considerable amount. Preparation for egress was conducted slowly, carefully, and deliberately, and future missions should be planned and conducted with the same philosophy. The extravehicular activity preparation checklist was adequate and was closely followed. However, minor items that required a decision in real time or had not been considered before flight required more time than anticipated.

***

Depressurization of the lunar module was one aspect of the mission that had never been completely performed on the ground. In the various altitude chamber tests of the spacecraft and the extravehicular mobility unit, a complete set of authentic conditions was never present. The depressurization of the lunar module through the bacteria filter took much longer than had been anticipated. The indicated cabin pressure did not go below 0.1 psi, and some concern was experienced in opening the forward hatch against this residual pressure. The hatch appeared to bend on initial opening, and small particles appeared to be blown out around the hatch when the seal was broken . . .

4.12.3 Lunar Module Egress

Simulation work in both the water immersion facility and the 1/6-g environment in an airplane was reasonably accurate in preparing the crew for lunar module egress. Body positioning and arching-the-back techniques that were required to exit the hatch were performed, and no unexpected problems were experienced. The forward platform was more than adequate to allow changing the body position from that used in egressing the hatch to that required for getting on the ladder. The first ladder step was somewhat difficult to see and required caution and forethought. In general, the hatch, porch, and ladder operation were not particularly difficult and caused little concern. Operations on the platform could be performed without losing body balance, and there was adequate room for maneuvering.

The initial operation of the lunar equipment conveyor in lowering the camera was satisfactory, but after the straps had become covered with lunar surface material, a problem arose in transporting the equipment back into the lunar module. Dust from this equipment fell back onto the lower crewmember and into the cabin and seemed to bind the conveyor so as to require considerable force to operate it. Alternatives in transporting equipment into the lunar module had been suggested before flight, and although no opportunity was available to evaluate these techniques, it is believed they might be an improvement over the conveyor.

4.12.4 Surface Exploration

Work in the 1/6-g environment was a pleasant experience. Adaptation to movement was not difficult and movement seemed to be natural. Certain specific peculiarities, such as the effect of the mass versus the lack of traction, can be anticipated but complete familiarization need not be pursued.

The most effective means of walking seemed to be the lope that evolved naturally. The fact that both feet were occasionally off the ground at the same time, plus the fact that the feet did not return to the surface as rapidly as on earth, required some anticipation before attempting to stop. Although movement was not difficult, there was noticeable resistance provided by the suit.

On future flights, crewmembers may want to consider kneeling in order to work with their hands. Getting to and from the kneeling position would be no problem, and being able to do more work with the hands would increase productive capability.

Photography with the Hasselblad cameras on the remote control unit mounts produced no problems. The first panorama was taken while the camera was hand-held; however, it was much easier to operate on the mount. The handle on the camera was adequate, and very few pictures were triggered inadvertently.

The solar wind experiment was easily deployed. As with the other operations involving lunar surface penetration, it was only possible to penetrate the lunar surface material only about 4 or 5 inches. The experiment mount was not quite as stable as desired, but it stayed erect.

The television system presented no difficulty except that the cord was continually in the way. At first, the white cord showed up well, but it soon became covered with dust and was therefore more difficult to see. The cable had a "set" from being coiled around the reel and it would not lie completely flat on the surface. Even when it was flat, however, a foot could still slide under it, and the Commander became entangled several times . . .

Collecting the bulk sample required more time than anticipated because the modular equipment stowage assembly table was in deep shadow, and collecting samples in that area was far less desirable than taking those in the sunlight. It was also desirable to take samples as far from the exhaust plume and propellant contamination as possible. An attempt was made to include a hard rock in each sample and approximately 20 trips were required to fill the box. As in simulations, the difficulty of scooping up the material without throwing it out as the scoop became free created some problem. It was almost impossible to collect a full scoop of material, and the task required about double the planned time.

Several of the operations would have been easier in sunlight. Although it was possible to see in the shadows, time must be allowed for dark adaptation when walking from the sunlight into shadow. On future missions, it would advantageous to conduct a yaw maneuver just prior to landing so that the descent stage work area would be in sunlight.

The scientific experiment package was easy to deploy manually, and some time was saved here. The package was easy to manage, but finding a level area was quite difficult. A good horizon reference was not available, and in the 1/6-g environment, physical cues were not as effective as in a one-g. Therefore, the selection of a deployment site for the experiments caused some problems. The experiments were placed in an area between shallow craters in surface material of the same consistency as the surrounding area and which should be stable. Considerable effort was required to change the slope of one of the experiments. It was not possible to lower the equipment by merely forcing it down, and it was necessary to move the experiment back and forth to scrape away the excess surface material.

No abnormal conditions were noted during the lunar module inspection. The insulation on the secondary struts had been damaged from the heat, but the primary struts were only singed or covered with soot. There was much less damage than on the examples that had been seen before flight.

Obtaining the core tube sample presented some difficulty. It was impossible to force the tube more than 4 or 5 inches into the surface material, yet the material provided insufficient resistance to hold the extension handle in the upright position. Since the handle had to be held upright, this precluded using both hands on the hammer. In addition, the resistance of the suit made it difficult to steady the core tube and swing with any great force. The hammer actually missed several times. Sufficient force was obtained to make dents in the handle, but the tube could be driven only to a depth of about 6 inches. Extraction offered little or virtually no resistance. Two samples were taken.

Insufficient time remained to take the documented sample, although as wide a variety of rocks was selected as remaining time permitted.

The performance of the extravehicular mobility unit was excellent. Neither crewman felt any thermal discomfort. The Commander used the minimum cooling mode for most of the surface operation. The Lunar Module Pilot switched to the maximum diverter valve position immediately after sublimator startup and operated at maximum position for 42 minutes before switching to the intermediate position. The switch remained in the intermediate position for the duration of the extravehicular activity. The thermal effect of shadowed areas versus those areas in sunlight was not detectable inside the suit.

The crewmen were kept physically cool and comfortable, and the ease of performing in the 1/6-g environment indicate that tasks requiring greater physical exertion may be undertaken on future flights. The Commander experienced some physical exertion while transporting the sample return container to the lunar module, but his physical limit had not been approached.

4.12.5 Lunar Module Ingress

Ingress to the lunar module produced no problems. The capability to do a vertical jump was used to an advantage in making the first step up the ladder. By doing a deep knee bend, then springing up the ladder, the Commander was able to guide his feet to the third step. Movements in the 1/6-g environment were slow enough to allow deliberate foot placement after the jump. The ladder was a bit slippery from the powdery surface material, but not dangerously so.

As previously stated, mobility on the platform was adequate for developing alternate methods of transferring equipment from the surface. The hatch opened easily, and the ingress technique developed before flight was satisfactory. A concerted effort to arch the back was required when about half way through the hatch, to keep the forward end of the portable life support system low enough to clear the hatch. There was very little exertion associated with transition to a standing position.

Because of the bulk of the extravehicular mobility unit, caution had to be exercised to avoid bumping into switches, circuit breakers, and other controls while moving around the cockpit. One circuit breaker was in fact broken as a result of contact . . .

Equipment jettison was performed as planned, and the time taken before flight in determining the items not required for lift-off was well spent. Considerable weight reduction and increase in space was realized. Discarding the equipment through the hatch was not difficult, and only one item remained on the platform. The post-ingress checklist procedures were performed without difficulty; the checklist was well planned and was followed precisely.

4.12.6 Lunar Rest Period

The rest period was almost a complete loss. The helmet and gloves were worn to relieve any subconscious anxiety about a loss of cabin pressure and presented no problem. But noise, lighting, and a lower-than-desired temperature were annoying. It was uncomfortably cool in the suits, even with the water-flow disconnected. Oxygen flow was finally cut off, and the helmets were removed, but the noise from the glycol pumps was then loud enough to interrupt sleep. The window shades did not completely block out light, and the cabin was illuminated by a combination of light through the shades, warning lights, and display lighting. The Commander rested on the ascent engine cover and was bothered by the light entering through the telescope. The Lunar Module Pilot estimated that he slept fitfully for perhaps 2 hours and the Commander did not sleep at all, even though body positioning was not a problem. Because of the reduced gravity, the positions on the floor and on the engine cover were both quite comfortable.

***

4.19 ENTRY

Because of the presence of thunderstorms in the primary recovery area (1285 miles downrange from the entry interface of 400,000 feet), the targeted landing point was moved to a range of 1500 miles from the entry interface. This change required the use of computer program P65 (skip-up control routine) in the computer, in addition to those programs used for the planned shorter range entry. This change caused the crew some apprehension, since such entries had rarely been practiced in preflight simulations. However, during the entry, these parameters remained within acceptable limits. The entry was guided automatically and was nominal in all respects. The first acceleration pulse reached approximately 6.5g and the second reached 6.0g.

4.20 RECOVERY

On the landing, the 18-knot surface wind filled the parachutes and immediately rotated the command module into the apex down (stable II) flotation position prior to parachute release. Moderate wave-induced oscillations accelerated the uprighting sequence, which was completed in less than 8 minutes. No difficulties were encountered in completing the postlanding checklist.

The biological isolation garments were donned inside the spacecraft. Crew transfer into the raft was followed by hatch closure and by decontamination of the spacecraft and crewmembers by germicidal scrubdown.

Helicopter pickup was performed as planned, but visibility was substantially degraded because of moisture condensation on the biological isolation garment faceplate. The helicopter transfer to the aircraft carrier was performed as quickly as could be expected, but the temperature increase inside the suit was uncomfortable. Transfer from the helicopter into the mobile quarantine facility completed the voyage of Apollo 11.

The goal set by John F. Kennedy just over eight years earlier had been met; Americans had flown to the Moon and returned safely to Earth. Apollo 11 was a success, technically and politically.

After Apollo 11, nine additional flights to the Moon, through

Apollo 20, had originally been planned. Apollo 12 through 15 would use the same basic equipment as had Apollo 11, but would land at different locations and stay for increasingly longer times on the lunar surface. Apollo 16 through 20 would carry a lunar rover, a small vehicle that would allow the astronauts to traverse the lunar surface. The lunar module would carry enough consumables (oxygen, food, etc.) to allow the crew to stay on the Moon for up to seventy-eight hours.

Apollo 12 was launched during a thunderstorm on November 14. Lightning struck the spacecraft during its initial ascent, and for a moment it appeared that the mission would have to be aborted. But this threat passed, and the lunar module made a precision landing within walking distance of the Surveyor 3 spacecraft that had landed on the moon in April 1967.

The next mission, Apollo 13, was launched on April 11, 1970. More than two days away from Earth on the mission's outbound journey, an oxygen tank in the service module exploded, placing the crew's life in jeopardy. This account by Apollo 13 commander Jim Lovell, written five years after the mission, captures the extraordinary efforts on the part of the crew and ground controllers to bring Lovell and his crewmates Fred Haise and Jack Swigert safely back to Earth.

James Lovell, "Houston, We've Had a Problem," NASA, *Apollo Expeditions to the Moon*, 1975

Since Apollo 13 many people have asked me, "Did you have suicide pills on board?" We didn't, and I never heard of such a thing in the eleven years I spent as an astronaut and NASA executive.

I did, of course, occasionally think of the possibility that the spacecraft explosion might maroon us in an enormous orbit about the Earth—a sort of perpetual monument to the space program. But Jack Swigert, Fred Haise, and I never talked about that fate during our perilous flight. I guess we were too busy struggling for survival.

Survive we did, but it was close. Our mission was a failure but I like to think it was a successful failure.

Apollo 13, scheduled to be the third lunar landing, was launched at 1313 Houston time on Saturday, April 11, 1970 . . .

Looking back, I realize I should have been alerted by several

omens that occurred in the final stages of the Apollo 13 preparation. First, our command module pilot, Ken Mattingly, with
whom Haise and I had trained for nearly two years, turned out
to have no immunity to German measles (a minor disease the
backup LM pilot, Charlie Duke, had inadvertently exposed us
to). I argued to keep Ken, who was one of the most conscientious, hardest working of all the astronauts. In my argument to
Dr. Paine, the NASA Administrator, I said, "Measles aren't that
bad, and if Ken came down with them, it would be on the way
home, which is a quiet part of the mission. From my experience as
command module pilot on Apollo 8, I know Fred and I could
bring the spacecraft home alone if we had to." Besides, I said, Ken
doesn't have the measles now, and he may never get them. (Five
years later, he still hadn't.)

Dr. Paine said no, the risk was too great. So I said in that case
we'll be happy to accept Jack Swigert, the backup CMP, a good
man (as indeed he proved to be, though he had only two days of
prime-crew training).

***

Then there was the No. 2 oxygen tank, serial number 10024X-
TA0009. This tank had been installed in the service module of
Apollo 10, but was removed for modification (and was damaged
in the process of removal) . . .

This tank was fixed, tested at the factory, installed in our
service module and tested again during the Countdown Demonstration Test at the Kennedy Space Center beginning March 16,
1970. The tanks normally are emptied to about half full, and
No. 1 behaved all right. But No. 2 dropped to only 92 percent of
capacity. Gaseous oxygen at 80 psi was applied through the vent
line to expel the liquid oxygen, but to no avail. An interim dis-
crepancy report was written, and on March 27, two weeks before
launch, detanking operations were resumed. No. 1 again emptied
normally, but its idiot twin did not. After a conference with contractor and NASA personnel, the test director decided to "boil
off" the remaining oxygen in No. 2 by using the electrical heater
within the tank. The technique worked, but it took eight hours of
65-volt DC power from the ground-support equipment to dissipate the oxygen.

With the wisdom of hindsight, I should have said, "Hold it.

Wait a second. I'm riding on this spacecraft. Just go out and replace that tank." But the truth is, I went along, and I must share the responsibility with many, many others for the $375 million failure of Apollo 13. On just about every spaceflight we have had some sort of failure, but in this case, it was an accumulation of human errors and technical anomalies that doomed Apollo 13 . . .

The first two days we ran into a couple of minor surprises, but generally Apollo 13 was looking like the smoothest flight of the program. At 46 hours 43 minutes Joe Kerwin, the CapCom on duty, said, "The spacecraft is in real good shape as far as we are concerned. We're bored to tears down here." It was the last time anyone would mention boredom for a long time.

At 55 hours 46 minutes, as we finished a 49-minute TV broadcast showing how comfortably we lived and worked in weightlessness, I pronounced the benediction: "This is the crew of Apollo 13 wishing everybody there a nice evening, and we're just about ready to close out our inspection of Aquarius (the LM) and get back for a pleasant evening in Odyssey (the CM). Good night."

On the tapes I sound mellow and benign, or some might say fat, dumb, and happy. A pleasant evening, indeed! Nine minutes later the roof fell in; rather, oxygen tank No. 2 blew up, causing No. 1 tank also to fail. We came to the slow conclusion that our normal supply of electricity, light, and water was lost, and we were about 200,000 miles from Earth. We did not even have power to gimbal the engine so we could begin an immediate return to Earth.

The message came in the form of a sharp bang and vibration. Jack Swigert saw a warning light that accompanied the bang, and said, "Houston, we've had a problem here." I came on and told the ground that it was a main B bus undervolt. The time was 2108 hours on April 13.

Next, the warning lights told us we had lost two of our three fuel cells, which were our prime source of electricity. Our first thoughts were ones of disappointment, since mission rules forbade a lunar landing with only one fuel cell.

With warning lights blinking on, I checked our situation; the quantity and pressure gages for the two oxygen tanks gave me cause for concern. One tank appeared to be completely empty, and there were indications that the oxygen in the second tank was rapidly being depleted. Were these just instrument malfunctions? I was soon to find out.

Thirteen minutes after the explosion, I happened to look out of the left-hand window, and saw the final evidence pointing toward potential catastrophe. "We are venting something out into the—into space," I reported to Houston. Jack Lousma, the CapCom replied, "Roger, we copy you venting." I said, "It's a gas of some sort."

It was a gas—oxygen—escaping at a high rate from our second, and last, oxygen tank. I am told that some amateur astronomers on top of a building in Houston could actually see the expanding sphere of gas around the spacecraft.

ARRANGING FOR SURVIVAL

The knot tightened in my stomach, and all regrets about not landing on the Moon vanished. Now it was strictly a case of survival. The first thing we did, even before we discovered the oxygen leak, was to try to close the hatch between the CM and the LM. We reacted spontaneously, like submarine crews, closing the hatches to limit the amount of flooding. First Jack and then I tried to lock the reluctant hatch, but the stubborn lid wouldn't stay shut! Exasperated, and realizing that we didn't have a cabin leak, we strapped the hatch to the CM couch. In retrospect, it was a good thing that we kept the tunnel open, because Fred and I would soon have to make a quick trip to the LM in our fight for survival. It is interesting to note that days later, just before we jettisoned the LM, when the hatch had to be closed and locked, Jack did it—easy as pie. That's the kind of flight it was.

The pressure in the No. 1 oxygen tank continued to drift downward; passing 300 psi, now heading toward 200 psi. Months later, after the accident investigation was complete, it was determined that, when No. 2 tank blew up, it either ruptured a line on the No. 1 tank, or caused one of the valves to leak. When the pressure reached 200 psi, it was obvious that we were going to lose all oxygen, which meant that the last fuel cell would also die. At 1 hour and 29 seconds after the bang, Jack Lousma, then CapCom, said after instructions from Flight Director Glynn Lunney: "It is slowly going to zero, and we are starting to think about the LM lifeboat." Swigert replied, "That's what we have been thinking about too."

A lot has been written about using the LM as a lifeboat after the CM has become disabled . . . Fred Haise, fortunately, held the reputation as the top astronaut expert on the LM—after spending fourteen months at the Grumman plant on Long Island, where the LM was built. Fred says: "I never heard of the LM being used in the sense that we used it. We had procedures, and we had trained to use it as a backup propulsion device, the rationale being that the thing we were really covering was the failure of the command module's main engine, the SPS engine. In that case, we would have used combinations of the LM descent engine, and in some cases, for some lunar aborts, the ascent engine as well. But we never really thought and planned, and obviously, we didn't have the procedures to cover a case where the command module would end up fully powered down."

To get Apollo 13 home would require a lot of innovation. Most of the material written about our mission describes the ground-based activities, and I certainly agree that without the splendid people in Mission Control, and their backups, we'd still be up there . . .

However, I would be remiss not to state that it really was the teamwork between the ground and flight crew that resulted in a successful return. I was blessed with two shipmates who were very knowledgeable about their spacecraft systems and the disabled service module forced me to relearn quickly how to control spacecraft attitude from the LM, a task that became more difficult when we turned off the attitude indicator.

With only 15 minutes of power left in the CM, CapCom told us to make our way into the LM. Fred and I quickly floated through the tunnel, leaving Jack to perform the last chores in our forlorn and pitiful CM that had seemed such a happy home less than two hours earlier. Fred said something that strikes me as funny as I read it now: "Didn't think I'd be back so soon." But nothing seemed funny in real time on that 13th of April, 1970.

There were many, many things to do. In the first place, did we have enough consumables to get home? Fred started calculating, keeping in mind that the LM was built for only a 45-hour lifetime, and we had to stretch that to 90 . . . It turned out that we had enough oxygen. The full LM descent tank alone would suffice, and in addition, there were two ascent-engine oxygen tanks, and two backpacks whose oxygen supply would never be used on the lunar surface. Two emergency bottles on top of those packs had six or seven pounds each in them. (At LM jettison, just before reentry, 28.5 pounds of oxygen remained, more than half of what we started with.)

We had 2181 ampere hours in the LM batteries. We thought that was enough if we turned off every electrical power device not absolutely necessary. We could not count on the precious CM batteries, because they would be needed for reentry after the LM was cast off. In fact, the ground carefully worked out a procedure where we charged the CM batteries with LM power. As it turned out, we reduced our energy consumption to a fifth of normal, which resulted in our having 20 percent of our LM electrical power left when we jettisoned Aquarius. We did have one electrical heart-stopper during the mission. One of the CM batteries vented with such force that it momentarily dropped off the line. We knew we were finished if we permanently lost that battery.

Water was the real problem. Fred figured that we would run out of water about five hours before we got back to Earth, which was calculated at around 151 hours. But even there, Fred had an ace in the hole. He knew we had a data point from Apollo 11, which had not sent its LM ascent stage crashing into the Moon, as subsequent missions did. An engineering test on the vehicle showed that its mechanisms could survive seven or eight hours in space without water cooling, until the guidance system rebelled at this enforced toasting. But we did conserve water. We cut down to six ounces each per day, a fifth of normal intake, and used fruit juices; we ate hot dogs and other wet-pack foods when we ate at all. (We lost hot water with the accident and dehydratable food is not palatable with cold water.) Somehow, one doesn't get very thirsty in space and we became quite dehydrated. I set one record that stood up throughout Apollo: I lost fourteen pounds, and our crew set another by losing a total of 31.5 pounds, nearly 50 percent more than any other crew. Those stringent measures resulted in our finishing with 28.2 pounds of water, about 9 percent of the total.

Fred had figured that we had enough lithium hydroxide canisters, which remove carbon dioxide from the spacecraft. There were four cartridge from the LM, and four from the backpacks, counting backups. But he forgot that there would be three of us in the LM instead of the normal two. The LM was designed to support two men for two days. Now it was being asked to care for three men nearly four days.

A SQUARE PEG IN A ROUND HOLE

We would have died of the exhaust from our own lungs if Mission Control hadn't come up with a marvelous fix. The trouble was the square lithium hydroxide canisters from the CM would not fit the round openings of those in the LM environmental system. After a day and a half in the LM a warning light showed us that the carbon dioxide had built up to a dangerous level, but the ground was ready. They had thought up a way to attach a CM canister to the LM system by using plastic bags, cardboard, and tape—all materials we had on board. Jack and I put it together: just like building a model airplane. The contraption wasn't very handsome, but it worked. It was a great improvisation—and a fine example of cooperation between ground and space.

The big question was, "How do we get back safely to Earth?" The LM navigation system wasn't designed to help us in this situation. Before the explosion, at 30 hours and 40 minutes, we had made the normal midcourse correction, which would take us out of a free-return-to-Earth trajectory and put us on our lunar landing course. Now we had to get back on that free-return course. The ground-computed 35-second burn, by an engine designed to land us on the Moon, accomplished that objective 5 hours after the explosion.

As we approached the Moon, the ground informed us that we would have to use the LM descent engine a second time; this time a long 5-minute burn to speed up our return home. The maneuver was to take place 2 hours after rounding the far side of the Moon, and I was busy running down the procedures we were to use. Suddenly, I noticed that Swigert and Haise had their cameras out and were busy photographing the lunar surface. I looked at them incredulously and said, "If we don't make this next maneuver correctly, you won't get your pictures developed!" They said, "Well, you've been here before and we haven't." Actually, some of the pictures these tourists took turned out to be very useful.

***

[After a successful alignment of the spacecraft guidance system using the Sun, rather than a star, as the reference point], I am told the cheer of the year went up in Mission Control. Flight Director Gerald Griffin, a man not easily shaken, recalls: "Some years later I went back to the log and looked up that mission. My writing was almost illegible I was so damned nervous. And I remember the exhilaration running through me: My God, that's kinda the last hurdle—if we can do that, I know we can make it. It was funny, because only the people involved knew how important it was to have that platform properly aligned." Yet Gerry Griffin barely mentioned the alignment in his change-of-shift briefing—"That check turned out real well" is all he said an hour after his penmanship failed him. Neither did we, as crew members, refer to it as a crisis in our press conference nor in later articles.

The alignment with the Sun proved to be less than a half a degree off. Hallelujah. Now we knew we could do the 5-minute P.C. + 2 burn with assurance, and that would cut the total time of our voyage to about 142 hours. We weren't exactly home free: we had a dead service module, a command module with no power, and a lunar module that was a wonderful vehicle to travel home in, but unfortunately didn't have a heat shield required to enter the Earth's atmosphere. But all we needed now was a continuation of the expertise we seemed blessed with, plus a little luck.

TIRED, HUNGRY, WET, COLD, DEHYDRATED

The trip was marked by discomfort beyond the lack of food and water. Sleep was almost impossible because of the cold. When we turned off the electrical systems, we lost our source of heat, and the Sun streaming in the windows didn't much help. We were as cold as frogs in a frozen pool, especially Jack Swigert, who got his feet wet and didn't have lunar overshoes. It wasn't simply that the temperature dropped to 38 F: the sight of perspiring walls and wet windows made it seem even colder. We considered putting on our spacesuits, but they would have been bulky and too sweaty. Our teflon-coated inflight coveralls were cold to the touch, and how we longed for some good old thermal underwear.

The ground, anxious not to disturb our homeward trajectory, told us not to dump any waste material overboard. What to do with urine taxed our ingenuity . . . I'm glad we got home when we did, because we were just about out of ideas for stowage.

A most remarkable achievement of Mission Control was quickly developing procedures for powering up the CM after its long cold sleep. They wrote the documents for this innovation in three days, instead of the usual three months. We found the CM a cold, clammy tin can when we started to power up. The walls, ceiling, floor, wire harnesses, and panels were all covered with droplets of water. We suspected conditions were the same behind the panels. The chances of short circuits caused us apprehension, to say the least. But thanks to the safeguards built into the command module after the disastrous fire in January 1967, no arcing took place. The droplets furnished one sensation as we decelerated in the atmosphere: it rained inside the CM.

Four hours before landing, we shed the service module; Mission Control had insisted on retaining it until then because everyone feared what the cold of space might do to the unsheltered CM heat shield. I'm glad we weren't able to see the SM earlier. With one whole panel missing, and wreckage hanging out, it was a sorry mess as it drifted away.

Three hours later we parted with faithful Aquarius, rather rudely, because we blasted it loose with pressure in the tunnel in order to make sure it completely cleared. Then we splashed down gently in the Pacific Ocean near Samoa, a beautiful landing in a blue-ink ocean on a lovely, lovely planet.

***

The Nixon administration was not interested in an ambitious post-Apollo program, and in early 1970 told NASA to plan for a much lower budget in coming years. This led to a series of NASA decisions on how best to adjust to a constrained budget outlook. George Low, who had become NASA deputy administrator in December 1969, announced on January 4, 1970, that NASA was canceling Apollo 20 and stretching out the remaining seven missions so that they would continue through 1974. Ten days later, faced with continuing budget cuts, administrator Thomas Paine announced that production of the Saturn V would be suspended indefinitely once the fifteenth vehicle had been completed. Within six months of the first landing on the Moon, the United States had essentially abandoned the heavy-lift capability that was the key to future space exploration.

More cancellations were to come. NASA leadership was faced with how to fit both the cost of future Apollo missions and getting started on its post-Apollo programs into the reduced budget proposed by the White House. In addition, some within NASA, especially after the near-tragedy of Apollo 13, judged that the benefits of additional Apollo missions were not worth the risks involved. Reluctantly, NASA leadership agreed that NASA should cancel Apollo 15, the last mission without a lunar roving vehicle, and Apollo 19. The remaining flights after Apollo 14 were then renumbered Apollo 15 through 17. The White House was formally notified of the NASA decision in a September 1, 1970, letter to President Nixon.

Letter to the President from T. O. Paine (Administrator, NASA), September 1, 1970

Dear Mr. President:

NASA has now completed an intensive review of the future of its manned flight program, including the six remaining Apollo lunar landing missions. We have determined that we should continue with four of the six previously planned lunar landings, but should drop the remaining two. It has been necessary to reach this decision at this time in order to avoid expenditures preparing for the Apollo 15 and 19 missions to be deleted, and to apply the necessary resources to those missions which are to be retained. During the conduct of this study, we requested and received the advice of the scientific community, and have kept your staff and the appropriate Congressional committees abreast of our effort.

The most compelling reason for the decision to delete these two flights, which we have arrived at reluctantly but with overwhelming consensus, is the current and reasonably foreseeable austere funding situation for NASA. As you know, we do not yet have our FY 1971 appropriation bill, but must proceed realistically to operate at a level not to exceed the $3.268.7 million voted by Congress in the vetoed HUD and Independent Offices Bill. This is $64 million below your budget request. In addition, a responsible and realistic view of the future indicates that increases in the NASA budget over the next few years will probably be less than required to carry out all of your first priority goals in aeronautics and space. We estimate that the deletion of these two lunar landing missions will reduce NASA's expenditures by $700–900 million over the next four years.

The scientific community, through our Lunar and Planetary Missions Board and the National Academy of Sciences' Space Science Board, strongly endorsed the retention of all six flights based on the scientific benefits to be derived. I fully understand and am sympathetic with their views, but these benefits do not, in our judgment, outweigh the benefits of other ongoing and future NASA programs and the risks involved in these difficult missions.

The six previously planned lunar flights would utilize all of the nation's available Saturn V launch vehicles, on which production was discontinued last year because of funding constraints. In view of Soviet progress on large launch vehicles, it is prudent to retain a modest Saturn V capability for such possible missions as launching a space station, retaining a back-up vehicle for scheduled missions, or for other unforeseen national requirements; deleting the Apollo 15 and 19 missions provides a national reserve of two Saturn V's.

You realize, I know, that each lunar mission is in itself a unique risk. We acknowledge this and must accept the responsibility. Obviously, there is an overall risk reduction in the dropping of two flights, and this advantage is multiplied by rescheduling the four remaining Apollo lunar landing missions to fly within a two-year period, without a significant gap between them. The momentum and morale engendered by such a compact schedule are distinct safety advantages in my view.

The proposed new manned flight program will fly the four remaining lunar flights at six-month intervals in 1971–72, followed by the Skylab program in which three teams of three astronauts will visit this space station prototype during 1973. There will then be a hiatus in the U.S. manned flight until the first test flights of the new space shuttle in 1975; these flights will continue through 1976, and the first manned orbital flight of the shuttle should be made in 1977. This program does represent a reduction in near-term manned flight, but it permits a sound development program for the space vehicle of the late 1970's. In light of the existing and foreseeable near-term funding situation, it is clearly the most realistic and forward-looking approach to the nation's future in space and aeronautics.

As early as December 1969, Richard Nixon had questioned the need for eight additional Apollo missions, and by early 1971 was pressuring his associates to cancel the last two Apollo missions, Apollo 16 and Apollo 17. The near-tragic Apollo 13 reinforced Nixon's skepticism regarding the wisdom of continued flights to the Moon. Thomas Paine resigned as NASA administrator in September 1970. George Low served as acting administrator until his successor could be chosen. James Fletcher, a former president of the University of Utah and an industrial executive, became NASA administrator in May 1971. In November of that year, he wrote the deputy director of the Office of Management and Budget, Caspar W. Weinberger, to provide both reasons why canceling the final two Apollo missions was not a good idea and a rationale and needed actions if the president persisted in his attempts to cancel the missions.

James C. Fletcher (Administrator, NASA), Letter to Caspar W. Weinberger (Deputy Director, Office of Management and Budget), November 3, 1971

Dear Cap:

In our conversation last week, you indicated that cancellation of Apollo 16 and 17 was being considered by the President and asked for my views on the actions that should be taken to offset or minimize the adverse consequences if such a decision is made.

From a scientific standpoint these final two missions are extremely important, especially Apollo 17 which will be the only flight carrying some of the most advanced experiments originally planned for Apollos 18 and 19, cancelled last year. With what we have learned from Apollo 15 and previous missions, we seem to be on the verge of discovering what the entire moon is like: its structure, its composition, its resources, and perhaps even its origin. If Apollo 16 and 17 lead to these discoveries, the Apollo program will go down in history not only as man's greatest adventure, but also as his greatest scientific achievement. Recognizing the great scientific potential and the relatively small saving ($133 million) compared to the investment already made in Apollo ($24 billion), I must as Administrator of NASA strongly recommend that the program be carried to completion as now planned.

If broader considerations, nevertheless, lead to a decision to cancel Apollo 16 and 17, the consequences would be much more serious than the loss of a major scientific opportunity. Unless compensatory actions are taken at the same time to offset and minimize the impact, this decision could be a blow from which the space program might not easily recover. As you requested, I will summarize the principal adverse consequences as I see them and then outline my recommendations on the compensatory actions necessary.

PRINCIPAL ADVERSE CONSEQUENCES

1. Negative Effect on Congressional and Public Support.

Without strong compensatory actions, a decision to cancel Apollo 16 and 17 would undermine the support the space program now enjoys and jeopardize the continued support that is required over the years to sustain the nation's position in space. Even though enthusiasm for the space program has diminished since the first lunar landing, NASA has continued to receive better than 98 percent of its budget requests each year (99.94% in FY 1972) because a substantial majority has accepted the judgments of the Administration and NASA's leadership that the space program is vital to the United States and that the programs recommended each year are necessary to achieve our goals. Cancellation of Apollo 16 and 17 would undermine this support in two ways.

First, it would call into question our credibility on this and other major elements of the space program since it would be a sudden reversal of the position we have so recently strongly supported in defense of our FY 1972 budget.

Second, it would terminate our best known, most visible and most exciting program which, in the minds of many in Congress and the public, has been the symbol of the space program and its success.

These factors, unless offset by strong positive actions, could result in a loss of confidence and interest that would have a "domino" effect, causing us to lose support for the programs which are essential to the long-term future of the nation in space.

2. Impact on Science and the Scientific Community.

At this time, the entire cognizant scientific community is strongly in favor of Apollo 16 and 17. Cancellation would come as a shock and a surprise in view of the strong support these missions have received from the President's Science Adviser, all of NASA's science advisory groups, NASA management, and the Congress. There will be strong and vocal critical reaction.

3. Impact on Industry.

Taken by itself, the direct impact of the cancellation of Apollo 16 and 17 would be further reductions in 1972 of over 6,000 aerospace jobs. The hardest hit areas would be Southern California, Long Island, Cape Kennedy, and Houston. Unless the decision is coupled with commitments and actions to proceed with and possibly expedite other programs, like the space shuttle, it will be a devastating blow, actually and psychologically, to an already hard-hit industry.

4. Impact on NASA.

The impact on NASA will be felt most strongly at Houston and, to a lesser extent, at Huntsville and Cape Kennedy. A major problem will be to hold together for over a year the team we will need to rely on to conduct safely the Skylab missions in 1973. We will have to deal with the difficult and visible problem of the futures of the 16 astronauts now assigned to Apollo 16 and 17. The blow to morale throughout NASA will be serious unless, again, the decision is coupled with clear decisions and commitments on future programs.

5. Impact on the Public.

The large segment of space enthusiasts in the population at large would be extremely disappointed by the proposed cancellation.

Included in this group would be millions who have come to Cape Kennedy, often from very long distances, to witness Apollo launches, and the much larger numbers who follow each mission closely on TV. These groups may be a minority in the U.S. but they are quite vocal and certainly non-negligible in size.

6. Impact Abroad.

It is our understanding from USIA reports that the Apollo flights have been a major plus factor for the U.S. image abroad. The impact of cancelling Apollo 16 and 17 should be assessed in arriving at a decision.

RATIONALE AND ACTIONS REQUIRED

If a decision is made to cancel Apollo 16 and 17, it is essential to provide a clearly stated and defensible rationale and take constructive actions to minimize the adverse impacts of the cancellation on the space program, the Administration, and the individuals, "communities of interest," and organizations affected. The rationale and actions must make it clear that, in spite of the cancellation, the President continues to support a program involving man in space and with strong scientific content.

***

Rationale

The rationale supporting this position would be as follows:

"Our space program has three basic purposes: exploration; the acquisition of scientific knowledge; and practical applications for man on earth. (See President's statement of March 7, 1970.) We must always strive to achieve the proper balance among these purposes.

"Today we must stress two aspects of our space program. We must give a top priority to practical applications now possible and press forward with the development of earth oriented systems which will enable us to make wider and more effective practical uses of space in the future.

"The key to the future in space—in science and exploration as well as practical applications—is routine access to space. Space activities will be part of our lives for the rest of time. These activities cannot continue, for long, to be as complex, as demanding, or as costly as they are today. We must develop new, simpler, less expensive techniques to go to space and to return from space. This is the goal of the space shuttle program. The sooner we get on with this development, the sooner will we be able to turn our knowledge gained in space science and space exploration toward helping man on earth.

"To operate in space most effectively we must also learn more about how man can best live and work in space. So while we are developing the shuttle, we must conduct space operations over longer periods of time—with Skylab.

"But to do all these things within limited resources, we must give up something. And when all factors are considered, the best project to give up—most reluctantly—is the remainder of Apollo: Apollo 16 and 17. This will for a time curtail our program of manned exploration and science.

"But we will, of course, continue exploration deep into space with unmanned spacecraft, including a landing on Mars in July 1976 with Viking, and the exploration of all the outer planets, Jupiter and beyond, with the Grand Tour late in this decade. The unmanned science program, with its High Energy Astronomy Observatory and other spacecraft will also continue to expand our fundamental knowledge of the universe. It is only manned science, and manned exploration, that will be curtailed.

"The United States must continue to fly men in space. Man will fly in space, and we cannot forego our responsibility—to ourselves and to the free world—to take part in this great venture. But for a time man can devote his own efforts, from space, toward practical needs here on earth, while leaving exploration beyond the earth to machines."

***

CONCLUSIONS

I recommend against the cancellation of Apollo 16 and 17 because these flights are scientifically important, and because much of the overall support for NASA's space program depends on our actions with respect to these flights.

If, nevertheless, for reasons external to NASA, Apollo 16 and 17 must be cancelled, then it becomes necessary to:

1. Provide strong backing to the manned earth-oriented space program.
2. Develop a rationale for the actions taken that is credible and supportable.
3. Take compensatory actions that will minimize the impact on the remaining NASA programs and their support.

The proposed rationale for the cancellation of Apollo 16 and 17 is that, in these times of pressing domestic needs, the manned space program should be earth-oriented instead of exploration and science-oriented.

The compensatory actions involve an early go-ahead for the space shuttle, the inclusion of "gap-filler" missions between Skylab and the shuttle, a number of augmented unmanned space science programs, and maintaining a total NASA budget at the FY 1971–1972 level of about $3.3 in budget authority.

Nixon's feelings on the two remaining Apollo missions were captured in a recording of a November 24, 1971, meeting discussing various budget issues among the president, his domestic policy adviser John Ehrlichman, his budget director George Shultz, and Secretary of the Treasury John Connally. Nixon did not want an Apollo launch to take place in the run-up to the November 1972 election. He was persuaded after this meeting to allow the Apollo 16 launch, which was moved up from its original launch date in mid-1972 to April 1972 so that it would not take place in close proximity to the election.

Recording of Oval Office Conversation, November 24, 1971

Nixon: Space, what's the problem here?

Ehrlichman: Well the problem here is do we go ahead with the next two shots [Apollo 16 and 17]?

Nixon: No! If we go, no shots before the election.

Ehrlichman: Then what would we do with all those employees?

Nixon: For those shots? How many, George?

Shultz: 17,500 or something like that.

Nixon: I don't like the feeling of space shots between now and the election.

Ehrlichman: But thinking of this thing [the space program] in just pure job terms, it is a hell of a job creator.

Connally: The American people are really not impressed by any more space shots.

Nixon: NASA is saying you'll find incredible things about the moon with these last two shots, and the American people say "so what?"

Shultz: Could I try another possibility? The last shot is the one in which they have loaded a great amount of scientific stuff from the ones that have been canceled before. That shot is scheduled after the election.

Nixon: I only see a minor waste of money. Keep the people on, but don't make the shots. I just don't feel the shots are a big deal at this time . . . There is also the risk you could have another Apollo 13 . . . That would be the worst thing we could have . . . We are just not going to do it. There will not be any launches between now and the election. The last shot, fine. Let's go forward with the last shot.

On August 13, 1971, NASA had announced that the crew for the last mission to the moon, Apollo 17, would include as lunar module pilot Harrison H. "Jack" Schmitt, a PhD geologist who had come to NASA as a scientist-astronaut in 1965 and had been deeply involved in planning the science to be done on the lunar missions. This assignment was the result of pressure from the scientific community, who argued that the Apollo missions should include at least one trained scientist as a crewmember.

Schmitt's selection came on the heels of the scientifically most successful mission to date, Apollo 15, launched on July 26, 1971. This was the first mission to carry the lunar roving vehicle, and astronauts David Scott and James Irwin used the vehicle to traverse almost seventeen miles of the lunar surface, a distance much greater than that traveled by the previous three crews. The crew spent three days on the Moon and conducted three exploration sorties using their lunar roving vehicle activities. Most significant, they identified and brought back to Earth specimens of the primitive lunar crust, the first material that had solidified from the molten outer layer of the young moon; one of these samples was dubbed the "Genesis rock."

The penultimate Apollo mission, Apollo 16, was launched on April 16, 1972. The mission landed in the lunar highlands, an area of the Moon that had not yet been explored. Apollo 16's objectives were similar to those of the preceding mission, with a focus on characterizing a region thought to be representative of much of the lunar surface.

All of the prior Apollo missions had been launched during daylight hours. After an almost three-hour delay, Apollo 17 lifted off at 12:33 a.m. EST on December 7, 1972. The vivid light from the Saturn V's five F-1 engines illuminated the night sky with an unreal brilliance. Astronauts Gene Cernan and Jack Schmitt spent three days exploring the lunar surface. As they prepared to leave the Moon, Cernan unveiled a plaque on the descent stage of the lunar module, which would remain on the Moon's surface. It read: "Here man completed his first explorations of the moon." As he took a last look at the lunar landscape, Cernan added, "As we leave the moon at Taurus-Littrow, we leave as we came, and, God willing, as we shall return, with peace and hope for mankind." The lunar module Challenger lifted off from the Moon at 5:55 p.m. EST on December 14.

With its departure, a remarkable era in human history came to a close. For the first time, human beings had left their home planet. In a statement issued after Apollo 17 had left the Moon, Richard Nixon declared, "This may be the last time in this century that men will walk on the Moon." With the decisions he made in the 1969–72 period, Nixon turned this prophecy into reality.

Richard Nixon, "Statement After Lift-Off from the Moon of Apollo 17 Lunar Module," December 14, 1972

As the Challenger leaves the surface of the Moon, we are conscious not of what we leave behind, but of what lies before us. The dreams that draw humanity forward seem always to be redeemed if we believe in them strongly enough and pursue them with diligence and courage. Once we stood mystified by the stars; today we reach out to them. We do this not only because it is man's destiny to dream the impossible, to dare the impossible, and to do the impossible, but also because in space, as on Earth, there are new answers and new opportunities for the improvement and the enlargement of human existence.

This may be the last time in this century that men will walk on the Moon. But space exploration will continue, the benefits of space exploration will continue, the search for knowledge through the exploration of space will continue, and there will be new dreams to pursue based on what we have learned. So let us neither mistake the significance nor miss the majesty of what we have witnessed. Few events have ever marked so clearly the passage of history from one epoch to another. If we understand this about the last flight of Apollo, then truly we shall have touched a "many-splendored thing."

To Gene Cernan, Jack Schmitt, and Ronald Evans, we say God speed you safely back to this good Earth.

CHAPTER 4

STEPS TOWARD
AN UNCERTAIN FUTURE

"Other than the long-range goal of sending humans to Mars, there is no strong, compelling national vision for the human spaceflight program. . . . The lack of national consensus on NASA's most publicly visible mission, along with out-year budget uncertainty, has resulted in the lack of strategic focus." This was the conclusion of a 2012 assessment of NASA's programs by the independent National Academy of Sciences. It came forty years after the last Apollo mission left the Moon and reflected the continuing absence in the years since Apollo of broadly accepted goals for the U.S. human spaceflight program and of a strategy to achieve them. This chapter will chart those years and, in doing so, attempt to explain the lack of agreement on the rationale and direction for the next steps in human spaceflight after Project Apollo. The included documents trace the decisions of the sequence of U.S. presidents who have collectively defined the tentative steps taken, and not taken, in the decades since humans last set foot on the Moon.

Of course, U.S. space activities have gone forward over the five decades since Americans last left the Moon. This chapter will also include documents highlighting space activities in that period, activities which resulted in significant accomplishments in low Earth orbit, as the space shuttle flew 133 successful missions, 37 of which were devoted to assembling and outfitting an orbiting research laboratory, the International Space Station. There was also tragedy, as two shuttle missions ended in catastrophe, each accident resulting in the deaths of a seven-person crew.

As he became president in January 1969, Richard Nixon recognized the need for a post-Apollo space strategy. The success of the December 1968 Apollo 8 mission had ensured that the first attempt at a lunar landing would come during 1969; that landing would achieve the goal set out by John F. Kennedy in 1961 that had guided the U.S. space effort during the 1960s. It was clear that decisions with respect to the goals and pace of the space program after the first lunar landing needed to be made, and that some sort of review would be the first step toward such decisions.

On February 13, 1969, Nixon asked his vice president, Spiro Agnew, to chair an interagency Space Task Group (STG) to provide a "definitive recommendation on the direction which the U.S. space program should take in the post Apollo period." This assignment went to Agnew in his role as chairman of the National Aeronautics and Space Council, not because he had any particular background on space matters. Nixon asked the STG to report to him by September 1.

Thomas Paine, who had become acting NASA administrator in November 1968, stayed on as the Nixon administration took office. Paine was a space visionary and an activist; he was not satisfied with being only a temporary caretaker, and in fact he hoped to be able to stay on in the permanent position even though he was a Democrat. By early 1969, NASA had identified a large, permanently occupied space station, supported by a low-cost logistics system christened the space shuttle, as its top-priority post-Apollo objective. Hoping to bypass the deliberations of the Space Task Group and to get an early endorsement of NASA's long-range ambitions, Paine went directly to the president with a carefully crafted case for such an action.

Memorandum from Thomas Paine (Administrator, NASA), for President Richard Nixon, February 26, 1969

. . . This memorandum outlines the problems, opportunities, and principal factors to be considered in Manned Space Flight, the area in our space program where NASA and your Administration are faced with the most urgent need for high-level decisions.

1. <u>Introduction</u>—NASA now has no approved plans or programs for manned space flight programs beyond the first Apollo manned lunar landings and the limited Apollo Applications earth orbital program now approved and

underway. Sharply reduced space budgets over the past three years and the failure of the previous Administration to make the required decisions and provide the necessary resources for future programs have built in a period of low accomplishment which will become apparent during your Administration, and have left the program without a clear sense of future direction for the post-Apollo period. Positive and timely action must be taken by your Administration now to prevent the nation's programs in manned space flight from slowing to a halt in 1972. The Apollo program served the nation well in providing a clear focus for the initial development and demonstration of manned space flight capabilities and technology. What is needed now, however, is a more balanced program for the next decade which will focus not on a single event but on sustained development and use of manned space flight over a period of years. As discussed below, there are two principal program opportunities: one is a long-term carefully-planned program of manned exploration of the moon, the other is a wide range of activities involved in the progressive development and operation of a permanent manned station in earth orbit. I believe that (a) manned lunar exploration should be continued at an economical rate to the point where a sound decision on the future course the nation should follow with respect to the moon can be made on the basis of knowledge and experience gained from a series of manned missions, and (b) the nation should, in any case, focus our manned space flight program for the next decade on the development and operation of a permanent space station—a National Research Center in earth orbit—accessible at reasonable cost to experts in many disciplines who can conduct investigations and operations in space which cannot be effectively carried out on earth.

2. **Status of U.S. Programs and Plans**—If our Apollo flights continue to be successful we will achieve the first manned lunar landing later this year, possibly as early as this summer. We will then carry out three additional landings at different locations on the moon, but the improved equipment required for moving beyond this with a scientifically significant lunar exploration plan is restricted to the study stage.

We will have a number of Saturn V boosters and Apollo spacecraft for future lunar missions left over from the Apollo program.

***

3. <u>USSR Prospects</u>—Recent USSR manned space flight activities substantiate previous indications that they are continuing strong programs pointed both at manned operations to the moon and at space station operations in earth orbit. Beyond this, they talk openly of future manned trips to the planets. While we now expect to land American astronauts on the moon before the Russians get there, the prospects are that during the period of our lunar flights in 1969–1970 the Soviets will, in addition to their manned lunar program, follow up their Soyuz 45 success by pushing toward a dominant position in large-scale long-duration space station operations in earth orbit. They will have the required heavy-lift launch capability. A multi-man, multi-purpose USSR space station operating in orbit before the U.S. could match it would give the USSR a strong advantage in space research and operations. Their moving clearly ahead of the U.S. in this field would have a continuing impact on the rest of the world, particularly if the U.S. program did not include a strong program in the earth orbital space station area.

4. <u>Opportunity for Leadership</u>—The fact that the previous Administration deferred to you the setting of the nation's goals in manned space flight creates a problem, but it also gives you a unique opportunity for leadership that will clearly identify your Administration with the establishment of the nation's major goals in manned space flight for the next decade. The impact and positive image of your leadership would be seriously downgraded in the eyes of the nation, the Congress, and the public, in my view, if the U.S. were once again placed in the position of reacting to Soviet initiatives in space. For this reason, I believe that you should consider the advisability of initiating a general directive to define the future goals of manned space flight in the next few months, prior to your final decisions on the plans that will be recommended to you on September 1 by the members of the Task Group you have established . . . I believe

that the case that a space station should be a major future
U.S. goal is now strong enough to justify at least a general
statement on your part that this will be one of our goals,
with the understanding (which could be reaffirmed in your
statement) that the scope, pace, specific uses, and detailed
plans of the space station will be determined on the basis of
the planning studies you have requested.

5. **Basic National Policy**—There is, I believe, almost unanimous agreement on the part of responsible leaders in your
Administration, the Congress, industry, the scientific community, and the general public that the U.S. must continue
manned space flight activities. The concerns and criticisms
that have been expressed do not question the continuation
of a manned space flight program but relate principally to
(a) the cost of the program, (b) the value of specific goals,
and (c) questions of priorities, within the space program or
between the space program and other scientific fields or
other national needs. However, virtually no responsible and
thoughtful person, to my knowledge, advocates or is prepared to accept the prospect of the United States abandon-
ing manned space flight to the Soviets to develop and exploit
as they see fit.

* * *

10. **Space Station**—With respect to future manned earth orbital
flight, the immediate problem is to assure that sufficient
funds are available in FY 1970 to permit detailed planning
and design studies to proceed, and to develop critical long
lead-time subsystems that will be required in any future
manned space flight program. Funds for these purposes
were specifically excluded from the present FY 1970 Budget, except for a small amount for studies, and we are therefore preparing an appropriate amendment to the FY 1970
Budget. This budget amendment can be approved now
without a commitment on your part to a permanent space
station as a major national goal. However, as stated in paragraph 4 above, we believe that it is in the national interest
for you to endorse this as a general U.S. objective at this
time. One possibility would be for you to give NASA and
the Task Group a specific instruction at the time you

approve the budget amendment that their recommendations to you in September should include proposals on the optimum program for establishing and utilizing a permanent U.S. space station.

11. **Space Station Concept**—The space station discussed here should become a central point for many activities in space and would be designed to carry on these activities in an effective and economic manner. It would be located in the most advantageous position to conduct investigations and operations in the space environment, many important aspects of which cannot be duplicated in an earth-based environment. The best place to study space is in space. We have in mind a system consisting of general and special-purpose modules with a low-cost logistic support system that will permit ready access and return by many users and their equipment and supplies. The space station would not be launched as a single unit, but would evolve over a period of years by adding to a core new modules as they are required and developed. One of the key objectives is to develop the system in cooperation with the Department of Defense so that it can be adaptable for future military research as well as for a variety of non-military scientific, engineering, and other application purposes.

There are many potential valuable uses of such a space station, and new ones will be found as experts in many fields become familiar with the possibilities and are able to visit and actually use it. However, we believe strongly that the justification for proceeding now with this major project as a national goal does not, and should not be made to depend on the specific contributions that can be foreseen today in particular scientific fields like astronomy or high energy physics, in particular economic applications, such as earth resources surveys, or in specific defense needs. Rather, the justification for the space station is that it is clearly the next major evolutionary step in man's experimentation, conquest, and use of space. The development of man's capability to live and work economically and effectively in space for long periods of time is an essential prerequisite not only for operations in earth orbit, but for long stay times on the moon and in the distant future, manned travel to the planets. It is for these reasons that I believe that space station development should become one of your Administration's principal working goals for the action over the next decade.

12. **Saturn V Production**—Under NASA's reduced 1969 operating plan and its present FY 1970 Budget, the production of Saturn V, the nation's largest launch vehicle, has been discontinued. The long-term future of the manned space flight program, as outlined above, will clearly require additional Saturn V launch vehicles, and we are therefore proposing a FY 1970 Budget amendment which will permit production to be resumed, at a very low rate, before "start up" costs become excessive. This amendment will not preclude other future decisions on large launch vehicles that might be made next fall, but it will assure that funds are available to provide the launch vehicles that will be needed. It will also get the U.S. out of what I believe to be a current untenable position of having discontinued production of our largest space booster at a time when the Soviets are expected to unveil a booster of this class or larger. For the reasons stated in paragraph 4 above, I recommend that you now take the initiative and announce this decision before the Russians launch their first booster in this class, so that your announcement will not be viewed as a reaction to the Soviet development.

13. **Cost**—In planning the space program careful consideration must, of course, be given each year, and especially at the time new major programs are undertaken, to the future budget levels required. Our national budget system wisely and necessarily provides for a review at least annually of both on-going and new programs, but long-term enterprises like major space programs require a policy commitment to follow through with the resources required over a period of many years . . .

Our present projections indicate that a balanced total NASA program that includes the recommended strong manned space flight program can be carried out with annual budgets over the next five years which will not rise above the $4.5 to $5.5 billion range. More precise projections will depend on the nature of the future lunar exploration and space station programs decided upon and on future decisions in areas other than manned space flight. By the time we submit the planning proposals to you in September we will be able to state with considerable confidence the projected future estimated costs of alternative total programs.

A total annual program level of $4.5–$5.5 billion compares to program and expenditure levels in the $5.0–$6.0 billion range reached in the 1964–1967 period, which in the past two years has been reduced to $3.9 billion in our FY 1969 operating plan and the present FY 1970 Budget . . .

This memorandum has given you my recommendation on the position your Administration should take with respect to the critical and urgent situation in manned space flight; other NASA problems and opportunities can be treated appropriately in the Task Group framework for your consideration in September. For the reasons stated above, and with the possibility of an initial lunar landing in July, I believe you should not defer initial consideration of the manned space flight problem. I therefore specifically recommend that you ask the members of the Task Group established in your memorandum of February 13, 1969, to meet within the next month and to consider as their first order of business the matters identified in this memorandum as requiring your early decision. They should then present their recommendations to you by the end of March. In anticipation of such a meeting, NASA will prepare and make available to the other members of the Task Group (a) detailed materials on the alternatives available, and (b) suggestions on how the recommended early decisions can be related to an effective process for developing overall space plans and alternatives for your consideration in September. I hope that this proposal will meet with your approval, and would, of course, be happy to discuss this matter further with you at your convenience.

The White House quickly rebuffed Paine's initiative, telling NASA that any decisions on future programs would have to await the recommendations of the Space Task Group. But during the summer of 1969, in the glow of the success of Apollo 11,

Paine and his associates George Mueller and Wernher von Braun were able to persuade the other members of the STG, especially Vice President Agnew, to put forward a very ambitious vision for NASA in the post-Apollo period. In its September 15, 1969, report, the Space Task Group set out several options for the future that included missions to Mars in the 1980s. The White House asked the STG to tone down these recommendations before submitting its report to the president, and so the introduction to the report proposed more modest commitments.

The report of the Space Task Group largely endorsed the hardware elements of the program that NASA had developed in the weeks following Apollo 11, including the development of a space station and a reusable transportation system. The Space Task Group as a whole did not recommend any particular option, though within a few days both Vice President Agnew and NASA Administrator Paine urged the president to approve Option II, which called for space station and space shuttle development by 1977 and an initial Mars expedition by 1986.

Space Task Group, "The Post-Apollo Space Program: Directions for the Future," September 1969

CONCLUSIONS AND RECOMMENDATIONS

The Space Task Group in its study of future directions in space, with recognition of the many achievements culminating in the successful flight of Apollo 11, views these achievements as only a beginning to the long-term exploration and use of space by man. We see a major role for this Nation in proceeding from the initial opening of this frontier to its exploitation for the benefit of mankind, and ultimately to the opening of new regions of space to access by man.

We have found increasing interest in the exploitation of our demonstrated space expertise and technology for the direct benefit of mankind in such areas as earth resources, communications, navigation, national security, science and technology, and international participation. We have concluded that the space program for the future must include increased emphasis upon space applications.

We have also found strong and wide-spread personal identification with the manned flight program, and with the outstanding men who have participated as astronauts in this program. We have concluded that a forward-looking space program for the future for this Nation should include continuation of manned space flight activity. Space will continue to provide new challenges to satisfy the innate desire of man to explore the limits of his reach.

We have surveyed the important national resource of skilled program managers, scientists, engineers, and workmen who have contributed so much to the success the space program has enjoyed. This resource together with industrial capabilities, government, and private facilities and growing expertise in space operations are the foundation upon which we can build.

We have found that this broad foundation has provided us with a wide variety of new and challenging opportunities from which to select our future directions. We have concluded that the Nation should seize these new opportunities, particularly to advance science and engineering, international relations, and enhance the prospects for peace.

We have found questions about national priorities, about the expense of manned flight operations, about new goals in space which could be interpreted as a "crash program." Principal concern in this area relates to decisions about a manned mission to Mars. We conclude that NASA has the demonstrated organizational competence and technology base, by virtue of the Apollo success and other achievements, to carry out a successful program to land man on Mars within 15 years. There are a number of precursor activities necessary before such a mission can be attempted. These activities can proceed without developments specific to a Manned Mars Mission—but for optimum benefit should be carried out with the Mars mission in mind. We conclude that a manned Mars mission should be accepted as a long-range goal for the space program. Acceptance of this goal would not give the manned Mars mission overriding priority relative to other program objectives, since options for decision on its specific date are inherent in a balanced program. Continuity of other unmanned exploration and applications efforts during periods of unusual budget constraints should be supported in all future plans.

We believe the Nation's future space program possesses potential for the following significant returns:

- new operational space applications to improve the quality of life on Earth
- non-provocative enhancement of our national security
- scientific and technological returns from space investments of the past decade and expansion of our understanding of the universe
- low-cost, flexible, long-lived, highly reliable, operational space systems with a high degree of commonality and reusability
- international involvement and participation on a broad basis.

Therefore, we recommend—

That this Nation accept the basic goal of a balanced manned and unmanned space program conducted for the benefit of all mankind.

To achieve this goal, the United States should emphasize the following program objectives:

- increase utilization of space capabilities for services to man, through an expanded space applications program
- enhance the defense posture of the United States and thereby support the broader objective of peace and security for the world through a program which exploits space techniques for accomplishment of military missions
- increase man's knowledge of the universe by conduct of a continuing strong program of lunar and planetary exploration, astronomy, physics, the earth and life sciences
- develop new systems and technology for space operations with emphasis upon the critical factors of: (1) commonality, (2) reusability, and (3) economy, through a program directed initially toward development of a new space transportation capability and space station modules which utilize this new capability
- promote a sense of world community through a program which provides opportunity for broad international participation and cooperation

As a focus for the development of new capability, we recommend the United States accept the long-range option or goal of manned planetary exploration with a manned Mars mission before the end of this century as the first target.

In proceeding towards this goal, three phases of activities can be identified:

- *initially*, activity should concentrate upon the dual theme of exploitation of existing capability and development of new capability, maintaining program balance within available resources.
- *second*, an operational phase in which new capability and new systems would be utilized in earth-moon space with groups of men living and working in this environment for extended periods of time. Continued exploitation of science and applications would be emphasized, making greater use of man or man-attendance as a result of anticipated lowered costs for these operations.
- *finally*, manned exploration missions out of earth-moon space, building upon the experience of the earlier two phases.

Schedule and budgetary implications associated with these three phases are subject to Presidential choice and decision at this time with detailed program elements to be determined in a normal annual budget and program review process. Should it be decided to develop concurrently the space transportation system and the modular space station, a rise of annual expenditures to approximately $6 billion in 1976 is required. A lower level of approximately $4–5 billion could be met if the space station and the transportation system were developed in series rather than in parallel . . .

The Space Task Group has had the opportunity to review the national space program at a particularly significant point in its evolution. We believe that the new directions we have identified can be both exciting and rewarding for this Nation. The environment in which the space program is viewed is a vibrant, changing one and the new opportunities that tomorrow will bring cannot be predicted with certainty. Our planning for the future should recognize this rapidly changing nature of opportunities in space . . .

THE POST-APOLLO SPACE PROGRAM: DIRECTIONS FOR THE FUTURE

I. INTRODUCTION

With the successful flight of Apollo 11, man took his first step on a heavenly body beyond his own planet. As we look into the distant future it seems clear that this is a milestone—a beginning—and not an end to the exploration and use of space. Success of the Apollo program has been the capstone to a series of significant accomplishments for the United States in space in a broad spectrum of manned and unmanned exploration missions and in the application of space techniques for the benefit of man. In the short span of twelve years man has suddenly opened an entirely new dimension for his activity. In addition, the national space program has made significant contributions to our national security, has been a political instrument of international value, has produced new science and technology, and has given us not only a national pride of accomplishment, but has offered a challenge and example for other national endeavors. The Nation now has the demonstrated capability to move on to new goals and new achievements in space in all of the areas pioneered during the decade of the sixties. In each area of space exploration what seemed impossible yesterday has become today's accomplishment. Our horizons and our competence have expanded to the point that we can consider unmanned missions to any region in our solar system; manned bases in earth orbit, lunar orbit or on the surface of the Moon; manned missions to Mars; space transportation systems that carry their payloads into orbit and then return and land as a conventional jet aircraft; reusable nuclear-powered rockets for space operations; remotely controlled roving science vehicles on the Moon or on Mars; and application of space capability to a variety of services of benefit to man here on earth. Our opportunities are great and we have a broad spectrum of choices available to us. It remains only to chart the course and to set the pace of progress in this new dimension for man.

***

III. GOALS AND OBJECTIVES

***

One of the values of the lunar landing goal was that it carried a definite time for its accomplishment, which stressed our technology and served as basis for planning and for budget support. It was a national commitment, a demonstration of the will and determination of the American people and of our technological competence at a time when these attributes were being questioned by many.

The need for an expression of our strength and determination as a Nation has changed considerably since that time. Today the need is for guidance—for direction—to set before the people a vision of where we are going . . .

***

In its deliberations, the Space Task Group considered a number of challenging new mission goals which were judged both technically feasible and achievable within a reasonable time, including establishment of a lunar orbit or surface base, a large 50–100 man earth-orbiting space base, and manned exploration of the planets. The Space Task Group believes that manned exploration of the planets is the most challenging and most comprehensive of the many long-range goals available to the Nation at this time, with manned exploration of Mars as the next step toward this goal. Manned planetary exploration would be a goal, not an immediate program commitment; it would constitute an understanding that within the context of a balanced space program, we will plan and move forward as a Nation towards the objective of a manned Mars landing before the end of this century. Mars is chosen because it is most earth-like, is in fairly close proximity to the Earth, and has the highest probability of supporting extraterrestrial life of all of the other planets in the solar system.

What are the implications of accepting this long-range goal or option on the character of the space program in the immediate future?

In a technical sense, the selection of manned exploration of the planets as a long-term option for the United States space program would act to focus a wide range of precursor activities and would be reflected in many decisions, large and small, where potential future applicability to long-lived manned planetary systems design will have relevance. In a broader sense such a selection would tend to reinforce and reaffirm the basic commitment to a long-term continued leadership position by the United States in space.

***

IV. PROGRAM AND BUDGET OPTIONS

***

The range of resource levels considered by the Task Group for NASA is shown in Figure 1.

These include: (1) an upper bound, defined by a program conducted at a maximum pace—limited, not by funds, but by technology; (2) options I, II, and III which illustrate programs consistent with the Task Group recommendations, but conducted under varying degrees of funding restraints; and (3) a low level program constructed with an increased unmanned science and applications effort consistent with the Task Group recommendations but, because of the significantly lower budget levels, without a manned flight program after completion of Apollo and Apollo Applications.

A comparison of the timing of major mission accomplishments under the various programs is indicated in Table 1. Although the program represented by the upper bound appears technically achievable, would provide maximum stimulation to our over-all capabilities, and is fully consistent with the Task Group recommendations, it represents on initial rate of growth of resources which cannot be realized because such budgetary requirements would substantially exceed predicted funding capabilities. This has therefore been rejected by the Space Task Group, and is presented only to demonstrate the upper bound of technological achievement.

***

The lower bound chosen by the Space Task Group illustrates a program conducted at significantly reduced funding levels. It is our judgment that, in order to achieve these significantly reduced NASA budgets, it would be necessary to reduce manned space flight operations below a viable minimum level. Therefore, this program has been constructed assuming a hiatus in manned flight following completion of Apollo applications and follow-on Apollo lunar missions. It thus sacrifices, for the period of such reduced budgets, program objectives relating to development of new capability, and the contribution of continuing manned space flight to several of the other program objectives recommended by the Task Group. It does, however, include a vigorous and expanded unmanned program of solar system exploration, astronomy, space applications for the benefit of man and potential for international cooperation. Funding for such a program would reduce gradually to a sustaining level of $2-3 billion depending upon the depth of change assumed for the supporting NASA facilities and manpower base.

The Space Task Group is convinced that a decision to phase out manned space flight operations, although painful, is the only way to achieve significant reductions in NASA budgets over the long term. At any level of mission activity, a continuing program of manned space flight, following use of launch vehicles and spacecraft purchased as part of Apollo, would require continued production of hardware, continued operation of extensive test, launch support and mission control facilities, and the maintenance of highly skilled teams of engineers, technicians, managers, and support personnel. Stretch-out of mission or production schedules, which can initially reduce total annual costs, would result in higher unit costs. More importantly, very low-level operations are highly wasteful of the skilled manpower required to carry out these operations and would risk deterioration of safety and reliability throughout the manned program. At some low level of activity, the viability at the program is in question. It is our belief that the interests of this Nation would not be served by a manned space flight program conducted at such levels.

＊

Table 1. Comparative Program Accomplishments

Milestones	Maximum Pace	Program I	II, III	Low Level
Manned Systems				
Space Station (Earth Orbit)	1975	1976	1977	—
50-Man Space Base (Earth Orbit)	1980	1980	1984	—
100-Man Space Base (Earth Orbit)	1985	1985	1989	—
Lunar Orbiting Station	1976	1978	1981	—
Lunar Surface Base	1978	1980	1983	—
Initial Mars Expedition	1981	1983	II–1986 III–Open	—
Space Transportation System				
Earth-to-Orbit	1975	1976	1977	—
Nuclear Orbit Transfer Stage	1978	1978	1981	—
Space Tug	1976	1978	1981	—
Scientific				
Large Orbiting Observatory	1979	1979	1980	—
High Energy Astron. Capability	1973	1973	1981	1973
Out-of-Ecliptic Survey	1975	1975	1978	1975
Mars—High Resolution Mapping	1977	1977	1981	1977
Venus—Atmospheric Probes	1976	1976	Mid-80s	1976
Multiple Outer Planet "Tours"	1977–79	1977–79	1977–79	1977–79
Asteroid Belt Survey	1975	1975	1981	1975
Applications				
Earliest Oper. Earth Resource System	1975	1975	1976	1975
Demonstration of Direct Broadcast	1978	1978	Mid-80s	1978
Demonstration of Navigation Traffic Control	1974	1974	1976	1974

These recommendations were not at all what Nixon and most of his advisers had in mind for the post-Apollo space effort; Nixon did not believe that there was public support for continuing a fast-paced space effort, and he gave higher priority to controlling the government budget than to investing in major new space initiatives. It took Nixon almost six months to issue a formal response to the Space Task Group report. In the interim there were cuts to the NASA budget, even after the successes of Apollo 11 and 12. In order to get money to study future programs such as the space station and space shuttle, NASA agreed to end production of the large Saturn V launch vehicle, thereby giving up the U.S. capability to send humans to the Moon and beyond. In this March 7, 1970, statement issued by the White House press office, Nixon set out his views on the principles that should guide decisions on future space efforts. Most significant, he indicated that the space program was no longer to be treated as a high-priority undertaking, but rather as just one of many things that the government would do, and that the space program would have to compete for priority and funding with other government programs. This "Nixon space doctrine" has been the guiding space policy for NASA in the years since 1970.

Richard M. Nixon, "Statement About the Future of the United States Space Program," March 7, 1970

Over the last decade, the principal goal of our Nation's space program has been the moon. By the end of that decade men from our planet had traveled to the moon on four occasions and twice they had walked on its surface. With these unforgettable experiences, we have gained a new perspective of ourselves and our world.

I believe these accomplishments should help us gain a new perspective of our space program as well. Having completed that long stride into the future which has been our objective for the past decade, we must now define new goals which make sense for the seventies. We must build on the successes of the past, always reaching out for new achievements. But we must also recognize that many critical problems here on this planet make high priority demands on our attention and our resources. By no means should we allow our space program to stagnate. But—with the entire future and the entire universe before us—we should not try to do everything at once. Our approach to space must continue to be bold—but it must also be balanced.

When this administration came into office, there were no clear, comprehensive plans for our space program after the first Apollo landing. To help remedy this situation, I established in February of 1969 a space task group, headed by the Vice President, to study possibilities for the future of that program. Their report was presented to me in September. After reviewing that report and considering our national priorities, I have reached a number of conclusions concerning the future pace and direction of the Nation's space efforts. The budget recommendations which I have sent to the Congress for fiscal year 1971 are based on these conclusions.

THREE GENERAL PURPOSES

In my judgment, three general purpose should guide our space program.

One purpose is exploration. From time immemorial, man has insisted on venturing into the unknown despite his inability to predict precisely the value of any given exploration. He has been willing to take risks, willing to be surprised, willing to adapt to new experiences. Man has come to feel that such quests are worthwhile in and of themselves—for they represent one way in which he expands his vision and expresses the human spirit. A great nation must always be an exploring nation if it wishes to remain great.

A second purpose of our space program is scientific knowledge—a greater systematic understanding about ourselves and our universe. With each of our space ventures, man's total information about nature has been dramatically expanded; the human race was able to learn more about the Moon and Mars in a few hours last summer than had been learned in all the centuries that had gone before. The people who perform this important work are not only those who walk in space suits while millions watch or those who launch powerful rockets in a burst of flame. Much of our scientific progress comes in laboratories and offices, where dedicated, inquiring men and women decipher new facts and add them to old ones in ways which reveal new truths. The abilities of these scientists constitute one of our most valuable national resources. I believe that our space program should help these people in their work and should be attentive to their suggestions.

A third purpose of the United States space effort is that of practical application—turning the lessons we learn in space to the early benefit of life on earth. Examples of such lessons are manifold; they range from new medical insights to new methods of communication, from better weather forecasts to new management techniques and new ways of providing energy. But these lessons will not apply themselves; we must make a concerted effort to see that the results of our space research are used to the maximum advantage of the human community.

A CONTINUING PROCESS

We must see our space effort, then, not only as an adventure of today but also as an investment in tomorrow. We did not go to the moon merely for the sport of it. To be sure, those undertakings have provided an exciting adventure for all mankind and we are proud that it was our Nation that met this challenge. But the most important thing about man's first footsteps on the moon is what they promise for the future.

We must realize that space activities will be a part of our lives for the rest of time. We must think of them as part of a continuing process—one which will go on day in and day out, year in and year out—and not as a series of separate leaps, each requiring a massive concentration of energy and will and accomplished on a crash timetable. Our space program should not be planned in a rigid manner, decade by decade, but on a continuing flexible basis, one which takes into account our changing needs and our expanding knowledge.

We must also realize that space expenditures must take their proper place within a rigorous system of national priorities. What we do in space from here on in must become a normal and regular part of our national life and must therefore be planned in conjunction with all of the other undertakings which are also important to us.

***

As we enter a new decade, we are conscious of the fact that man is also entering a new historic era. For the first time, he has reached beyond his planet; for the rest of time, we will think of ourselves as men from the planet earth. It is my hope that as we go forward with our space program, we can plan and work in a way which makes us proud both of the planet from which we come and of our ability to travel beyond it.

Discouraged by warnings of constrained budgets from Nixon and the Office of Management and Budget, Thomas Paine left NASA in September 1970, and his deputy, George Low, was designated acting administrator. The first program that NASA had hoped to gain post-Apollo approval to develop was a twelve-person space station launched by the large rocket developed for Apollo, the Saturn V. In addition, NASA had proposed developing the space shuttle as a fully reusable Earth-to-orbit launch vehicle to carry crew and supplies to the space station and to carry out other space operations. When the Nixon administration as part of its decision to reduce the NASA budget ended Saturn V production and refused to approve the space station, NASA in the fall of 1970 revisited its plans. The space agency turned its attention to gaining White House approval for the space shuttle as NASA's centerpiece post-Apollo program. To do so, NASA had to present a new rationale for the shuttle, as a cost-effective way of launching all U.S. civilian and national security payloads and providing new capabilities for space operations. It did not make sense to develop the space station without a shuttle, but it was possible to create a rationale for developing a space shuttle without tying it to the space station.

These two letters from George Low reflect a fundamental reshaping of the post-Apollo human spaceflight program. NASA also knew that President Nixon wanted to continue U.S. human spaceflight. In his October 28 letter to OMB's Caspar Weinberger, Low pointed out that by approving shuttle development, the United States could have a continuing human spaceflight program without setting a major new space goal. This proved to be a winning argument.

Letter from George Low (Acting NASA Administrator), to George Shultz (Director, Office of Management and Budget), September 30, 1970

Dear Mr. Shultz:

The purpose of this letter is to transmit the FY 1972 budget recommendations for the National Aeronautics and Space Administration and to set forth the major considerations which influenced their development.

***

5. In recognition of current and longer term fiscal restraints, we have curtailed existing FY 1971 programs and adjusted the time-phasing of the future programs to avoid commitments to excessive funding levels in 1972 and future years. Our recent decisions to cancel two Apollo flights result in substantially lower expenditures than previously projected through FY 1974. Our decision not to proceed with the simultaneous development of the space station and space shuttle, and to defer until later in the 1970's the actual development of the post-Skylab space station, removes one of the principal causes of an unacceptable peaking of the NASA budget at over $5 billion in the middle 1970's. As a result of these and other actions, we are now able to present and recommend an FY 1972 program which enables us to move toward each of the objectives in the President's statement without committing the nation to an annual NASA budget level in excess of $4 billion.

6. The key element in our program for the 1970's is the space shuttle. It supports the last four of the President's six objectives. We must start this development now to lay the foundations for the nation's future space program, and to bring about the major economies in later years. This can be done within the fiscal limits previously discussed.

***

3. <u>Reduce substantially the cost of space operations</u>—The space shuttle will be used for manned and man-tended experiments and to place unmanned scientific, weather, earth resources, and other satellites in earth orbit and bring them

back to earth for repair and reuse. In the future the space shuttle will also transport men, supplies, and scientific equipment to and from space stations. Through reusability, the space shuttle system will have a recurring operating cost per mission substantially lower than the cost of current systems. Of even greater significance from a cost standpoint, major reductions will be possible in the cost of scientific and applications programs because of the relaxation of size and weight constraints and the capability for recovery, repair and reuse of payloads.

We will be ready to proceed in FY 1972 on a realistic schedule with detailed design and initial development efforts which we estimate will lead to the first horizontal test flights in 1975. The first vertical flight would take place in 1977 with a manned orbital flight test possible by the end of that year, leading to an initial operating capability in the 1978/1979 time period.

We place the highest importance on proceeding expeditiously with this program. Not only is it important to minimize the gap in U.S. manned space flights following Skylab in 1972–73, but economic analyses, which strongly support the development of the space shuttle on a cost-effective basis, show that a delay of a year in its availability development could increase the total cost of the U.S. space program by as much as $2 billion.

4. **Extend man's capability to live and work in space**—The Skylab project, now in advanced stages of development, which requires substantial funding in FY 1972, is our only present flight program directed at this goal. Skylab will extend man's exposure to the space environment to 56 days, will perform an important manned solar astronomy experiment, and will extend our earth resources experiments beyond those carried out in the unmanned ERTS [Earth Resources Technology Satellite] program. After its launch in late 1972 and three revisit missions through the first half of 1973, no further manned missions using Apollo hardware are planned. We have made a major decision to defer development of a space station or "Skylab II" to a later time and to orient the space station studies we will continue in

FY 1972 toward modular systems that can be launched as well as serviced by the space shuttle.

* * *

Letter from George Low (Acting NASA Administrator), to Caspar Weinberger (Deputy Director, Office of Management and Budget), October 28, 1970

Dear Mr. Weinberger:

The discussions that NASA has had during the past weeks with OMB have covered, I believe, most of the matters of concern to us with regard to our FY 1972 budget. However, there are a few points—stated below in brief outline form—which I want to be especially sure are clearly understood and given careful consideration in the final decision process on the NASA FY 1972 budget.

* * *

2. Space Shuttle

- Our technical and management judgment is that we are progressing well on a realistic schedule of design studies and supporting technology development for the space shuttle, and will be ready to move in FY 1972 with detailed design and initial development as proposed in our FY 1972 recommendations.
- To support this schedule on an orderly, non-crash basis, the full amount ($190 million) of FY 1972 estimates for the shuttle is required.
- The shuttle is clearly the key to effective and economical use of space in the future:
    - It will provide a single improved and more economical system for unmanned and manned missions of the future.
    - Its development can be justified as a versatile and economical system for placing <u>unmanned</u> civil and military satellites in orbit, entirely apart from its role in conducting or supporting manned missions.
- A <u>clear go-ahead</u> to the space shuttle system should be given at this time because:

- The U.S. should not have a national posture which leaves the field of manned space flight to the USSR after Skylab.
- With the shuttle, the U.S. can have a continuing program of manned space flight without maintaining, after 1973, the expensive base required for Apollo/Skylab operations and without a commitment to a major new manned mission goal.
- An FY 1972 go-ahead will already leave a significant gap in U.S. manned space operations. Any delay will only widen this gap
- Proceeding with the shuttle now will be a significant step in halting further erosion of U.S. aerospace capability and in offsetting the adverse impacts of current space and defense cutbacks on the U.S. aerospace industry, both of which are matters of serious concern
- A clear sign that the U.S. is proceeding with the shuttle is needed to support our efforts to secure substantial international participation; without it prospects for European participation—now surprisingly good—will evaporate.

***

The forgoing points, together with those made in my letter of September 30, require, in my view, careful consideration. We will, of course, be glad to provide any further information you require.

In 1970, NASA did not obtain the shuttle approval it sought. That approval needed to come in 1971, the agency argued, or its human spaceflight program would fade away after the 1973 launch of the Apollo-derived Skylab space station. A new NASA administrator, James Fletcher, arrived in May 1971. That summer, NASA and the White House were locked in a bitter battle over the agency's future programs and budgets. At issue was whether NASA would receive approval to continue the Apollo program through the Apollo 16 and 17 missions and to begin developing the space shuttle. Fletcher and Low, who stayed on as deputy administrator after Fletcher's arrival, soon decided to abandon a fully reusable shuttle design as both too expensive

*and technically too challenging. The question then became what
shuttle design NASA would propose.*

*In the second half of 1971, NASA scrambled to select a design
that maintained the original shuttle's capabilities but could be
developed on a substantially lower budget. Leading the skepticism regarding the wisdom of going ahead with the shuttle was
the OMB staff. But OMB deputy director Weinberger disagreed
with his staff; he decided that budget cutting was going too far.
He told President Nixon in this August 1971 memorandum that
not approving the shuttle and canceling the two remaining
Apollo missions would be a mistake. Nixon wrote on the memo,
"I agree with Cap." This exchange between Weinberger and the
president was not reported to the OMB staff, who continued
through the rest of 1971 to oppose NASA's shuttle proposal and
to propose further reductions in the NASA budget.*

Caspar W. Weinberger (Deputy Director, Office of Management and Budget), via George P. Shultz, Memorandum for the President, "Future of NASA," August 12, 1971

Present tentative plans call for major reductions or change in
NASA, by eliminating the last two Apollo flights (16 and 17), and
eliminating or sharply reducing the balance of the Manned Space
Program (Skylab and Space Shuttle) and many remaining NASA
programs.

I believe this would be a mistake.

1. The real reason for sharp reductions in the NASA budget is
    that NASA is entirely in the 28% of the budget that is con-
    trollable. In short we cut it because it is cuttable, not be-
    cause it is doing a bad job or an unnecessary one.

2. We are being driven, by the uncontrollable items, to spend
    more and more on programs that offer no real hope for the
    future: Model Cities, OEO, Welfare, interest on National
    Debt, unemployment compensation, Medicare, etc. Of
    course, some of these have to be continued, in one form or
    another, but essentially they are programs, not of our
    choice, designed to repair mistakes of the past, not of our
    making.

3. We do need to reduce the budget, in my opinion, but we
    should not make all our reduction decisions on the basis of

what is reducible, rather than on the merits of individual programs.

4. There is real merit to the future of NASA, and to its proposed programs. The Space Shuttle and NERVA [an experimental nuclear-powered rocket engine] particularly offer the opportunity, among other things, to secure substantial scientific fall-out for the civilian economy at the same time that large numbers of valuable (and hard-to-employ-elsewhere) scientists and technicians are kept at work on projects that increase our knowledge of space, our ability to develop for lower cost space exploration, travel, and to secure, through NERVA, twice the existing propulsion efficiency of our rockets.

    It is very difficult to re-assemble the NASA teams should it be decided later, after major stoppages, to re-start some of the long-range programs.

5. Recent Apollo flights have been very successful from all points of view. Most important is the fact that they give the American people a much needed lift in spirit, (and the people of the world an equally needed look at American superiority). Announcement now, or very shortly, that we were cancelling Apollo 16 and 17 (an announcement we would have to make very soon if any real savings are to be realized) would have a very bad effect, coming so soon after Apollo 15's triumph. It would be confirming, in some respects, a belief that I fear is gaining credence at home and abroad: That our best years are behind us, that we are turning inward, reducing our defense commitments, and voluntarily starting to give up our super-power status, and our desire to maintain world superiority.

    America should be able to afford something besides increased welfare, programs to repair our cities, or Appalachian relief and the like.

6. I do not propose that we necessarily fund all NASA seeks—only that we couple any announcement to that effect with announcements that we are going to fund space shuttles, NERVA, or other major, future NASA activities . . .

Throughout the fall of 1971 suggestions for alternative shuttle designs emerged from NASA, its contractors, an advisory panel set up by presidential science adviser Ed David, and even from the contractor, Mathematica, that OMB had forced NASA to hire to carry out an analysis of shuttle economics. By November, OMB had developed, with the help of aerospace industry contractors it would not identify, a preferred shuttle design, a small vehicle with limited capabilities intended to test various aspects of shuttle re-usability and capabilities. NASA settled on a larger, "full capability" version of the shuttle, with a large disposable external fuel tank and strap-on boosters to provide the initial thrust to lift the shuttle orbiter off the launchpad. On November 22, as a final decision on the shuttle neared, NASA administrator Fletcher made a "best case" argument for the NASA design to the White House.

James C. Fletcher, "The Space Shuttle," November 22, 1971

This paper outlines NASA's case for proceeding with the space shuttle. The principal points are as follows:

1. The U.S. cannot forego manned space flight.
2. The space shuttle is the only meaningful new manned space program that can be accomplished on a modest budget.
3. The space shuttle is a necessary next step for the practical use of space. It will help
    * space science,
    * civilian space applications,
    * military space applications, and
    * the U.S. position in international competition and cooperation in space.
4. The cost and complexity of today's shuttle is *one-half* of what it was six months ago.
5. Starting the shuttle now will have significant positive effect on aerospace employment. Not starting would be a serious blow to both the morale and health of the Aerospace Industry.

THE U.S. CANNOT FOREGO MANNED SPACE FLIGHT

Man has worked hard to achieve—and has indeed achieved—the freedom of mobility on land, the freedom of sailing on his oceans, and the freedom of flying in the atmosphere.

And now, within the last dozen years, man has discovered that he can also have the freedom of space. Russians and Americans, at almost the same time, first took tentative small steps beyond the earth's atmosphere, and soon learned to operate, to maneuver, and to rendezvous and dock in near-earth space. Americans went on to set foot on the moon, while the Russians have continued to expand their capabilities in near-earth space. Man has learned to fly in space, and man will continue to fly in space. This is fact. And, given this fact, the United States cannot forego its responsibility—to itself and to the free world—to have a part in manned space flight. Space is not all remote. Men in near-earth orbit can be less than 100 miles from any point on earth—no farther from the U.S. than Cuba. For the U.S. not to be in space, while others do have men in space, is unthinkable, and a position which America cannot accept.

WHY THE SPACE SHUTTLE?

There are three reasons why the space shuttle is the right next step in manned space flight and the U.S. space program:

First, the shuttle is the only meaningful new manned space program which can be accomplished on a modest budget. Somewhat less expensive "space acrobatics" programs can be imagined but would accomplish little and be dead-ended. Additional Apollo or Skylab flights would be very costly, especially as left-over Apollo components run out, and would give diminishing returns. Meaningful alternatives, such as a space laboratory or a revisit to the moon to establish semi-permanent bases are *much* more expensive, and a visit to Mars, although exciting and interesting, is completely beyond our means at the present time.

Second, the space shuttle is needed to make space operations less complex and less costly. Today we have to mount an enormous effort every time we launch a manned vehicle, or even a large unmanned mission. The reusable space shuttle gives us a way to avoid this. This airplane-like spacecraft will make a launch into orbit an almost routine event—at a cost 1/10th of today's cost of space operations. How is this possible? Simply by not throwing everything away after we have used it just once—just as we don't throw away an airplane after its first trip from Washington to Los Angeles.

The shuttle even looks like an airplane, but it has rocket engines instead of jet engines. It is launched vertically, flies into orbit under its own power, stays there as long as it is needed, then glides back into the atmosphere and lands on a runway, ready for its next use. And it will do this so economically that, if necessary, it can provide transportation to and from space each week, at an annual operating cost that is equivalent to only 15 percent of today's total NASA budget, or about the total cost of a single Apollo flight. Space operations would indeed become *routine*.

Third, the space shuttle is needed to do useful things. The long term need is clear. In the 1980's and beyond, the low cost to orbit the shuttle gives is essential for all the dramatic and practical future programs we can conceive. One example is a space station. Such a system would allow many men to spend long periods engaged in scientific, military, or even commercial activities in a more or less permanent station which could be visited cheaply and frequently and refurbished, by means of a shuttle. Another interesting example is revisits to the moon to establish bases there; the shuttle would take the systems needed to earth orbit for assembly.

But what will the shuttle do before then? Why are routine operations so important? There is no single answer to these questions as there are many areas—in science, in civilian applications, and in military applications—where we can see now that the shuttle is needed; and there will be many more by the time routine shuttle services are actually available.

Take, for example, *space science*. Today it takes two to five years to get a new experiment ready for space flight, simply because operations in space are so costly that extreme care is taken to make everything just right. And because it takes so long, many investigations that should be carried out—to get fundamental knowledge about the sun, the stars, the universe, and, therefore, about ourselves on earth—are just not undertaken. At the same time, we have already demonstrated, by taking scientists and their instruments up in a Convair 990 airplane, that space science can be done in a much more straight-forward way with a much smaller investment in time and money, and with an ability to react quickly to new discoveries, because airplane operations are *routine*. This is what the shuttle will do for space science.

Or take *civilian space applications*. Today new experiments in space communications, or in earth resources, are difficult and expensive for the same reasons as discussed under science. But with routine space operations instruments could quickly be adjusted until the optimum combination is found for any given application—a process that today involves several satellites, several years of time, and great expense.

One can also imagine new applications that would only be feasible with the routine operation of the space shuttle. For example, it may prove possible (with an economical space transportation system, such as the shuttle) to place into orbit huge fields of solar batteries—and then beam the collected energy down to earth. This would be a truly pollution-free power source that does not require the earth's latent energy sources. Or perhaps one could develop a global environmental monitoring system, international in scope, that could help control the mess man has made of our environment. These are just two examples of what might be done with *routine* space shuttle operations.

What about *military space applications*? It is true that our military planning has not yet defined a specific need for man in space for military purposes. But will this always be the case? Have the Russians made the same decision? If not, the shuttle will be there to provide, quickly and routinely, for military operations in space, whatever they may be. It will give us a quick reaction time and the ability to fly ad hoc military missions whenever they are necessary. In any event, even without *new* military needs, the shuttle will provide the transportation for today's rocket-launched military spacecraft at substantially reduced cost.

Finally, the shuttle helps our *international* position—both our *competitive* position with the Soviets and our prospects of *cooperation* with them and with other nations.

Without the shuttle, when our present manned space program ends in 1973 we will surrender center stage in space to the only other nation that has the determination and capability to occupy it. The United States and the whole free world would then face a decade or more in which Soviet supremacy in space would be unchallenged. With the shuttle, the United States will have a clear space superiority over the rest of the world because of the low cost to orbit and the inherent flexibility and quick reaction capability of a reusable system.

The rest of the world—the free world at least—would depend on the United States for launch of most of their payloads.

On the side of cooperation, the shuttle would encourage far greater international participation in space flight. Scientists—as well as astronauts—of many nations could be taken along, with their own experiments, because shuttle operations will be routine. We are already discussing compatible docking systems with the Soviets, so that their spacecraft and ours can join in space. Perhaps ultimately men of all nations will work together in space—in joint environmental monitoring, international disarmament inspections, or perhaps even in joint commercial enterprises—and through these activities help humanity work together better on its planet earth. Is there a more hopeful way?

THE COST OF THE SHUTTLE
HAS BEEN CUT IN HALF

Six months ago NASA's plan for the shuttle was one involving heavy investment—$10 billion before the first manned orbital flight—in order to achieve a very low subsequent cost per flight—less than $5 million. But since then the design has been refined, and a trade-off has been made between investment cost and operational cost per flight. The result: a shuttle that can be developed for an investment of $4.5–$5 billion over a period of six years that will still only cost around $10 million or less per flight. (This means 30 flights per year at an annual cost for space transportation of 10 percent of today's NASA total budget, or one flight per week for 15 percent.) This reduction in investment cost was partly the result of a trade-off just mentioned, and partly due to a series of technical changes. The orbiter has been drastically reduced in size—from a length of 206 feet down to 110 feet. But the payload carrying capability has not been reduced: it is still 40,000 lbs. in polar orbit, or 65,000 lbs. in an easterly orbit, in a payload compartment that measures 15x60 feet.

The reduction in investment cost is highly significant. It means that the peak funding requirements, in any one year, can be kept down to a level that, even in a highly constrained NASA budget, will still allow for major advances in space science and applications, as well as in aeronautics.

THE SHUTTLE AND THE AEROSPACE INDUSTRY

The shuttle is a technological challenge requiring the kind of capability that exists today in the aerospace industry. An accelerated start on the shuttle would lead to a direct employment of 8,800 by the end of 1972, and 24,000 by the end of 1973. This cannot compensate for the 270,000 laid off by NASA cutbacks since the peak of the Apollo program but would take up the slack of further layoffs from Skylab and the remainder of the Apollo programs.

CONCLUSIONS

Given the fact that manned space flight is part of our lives, and that the U.S. must take part in it, it is essential to reduce drastically the complexity and cost of manned space operations. Only the space shuttle will do this. It will provide both *routine* and *quick reaction* space operations for space science and for civilian and military applications. The shuttle will do this at an investment cost that fits well within a highly constrained NASA budget. It will have low operating costs, and allow 30 to 50 space flights per year at a transportation cost equivalent to 10–15 percent of today's total NASA budget.

Two days later, Richard Nixon discussed the shuttle decision with his top policy adviser, John Ehrlichman, and OMB director George Shultz. It turned out that a major consideration in Nixon's mind was the employment impact of shuttle approval, particularly in California, a state key to Nixon's hope for reelection in November 1972. Shultz pointed out to Nixon that there was a more modest option than the full-capability shuttle that NASA was proposing, and Nixon indicated that he preferred the more modest option.

Recording of Oval Office Conversation, November 24, 1971

Ehrlichman: The Southern California people have a mighty press on for the space shuttle to be located in southern California. It is a highly visible kind of thing, if we were to announce at the

State of the Union or sometime that you were going ahead with the shuttle.

Nixon: This is not a State of the Union thing. I should do it [announcing shuttle approval] out in California where you are going to put it. Jobs—right, John? Do it in terms of jobs. It ought to be in California.

Shultz: NASA has a full thrust [shuttle] program, but there are options that are a little more modest.

Nixon: Take the more modest option. We'll take a look later to see [if that is the right choice]. It's the symbol that we are going to go forward. We are going to be positive on space. Nobody is going to be against us if we go forward in space, and a few will be for us because we do.

Ehrlichman: If you tell the aerospace industry that we are going ahead on the shuttle, that helps right now.

Shultz: While the shuttle and Skylab will keep men in space to a degree, the direction of this program ought to shift away from man in space and toward doing most of these things on an unmanned basis.

Nixon: I agree. Manned space flight becomes a stunt after a while.

Nixon's preference for a more modest version of the shuttle set up a heated conflict in December between OMB and NASA over what shuttle design to approve. Under continuing pressure from the OMB to lower the development costs of the shuttle, NASA in late December 1971 reluctantly changed its recommended configuration to one with a smaller payload capacity. In this letter, NASA administrator James Fletcher made what he believed to be NASA's final arguments for shuttle approval.

James C. Fletcher (Administrator, NASA), to Caspar W. Weinberger (Deputy Director, Office of Management and Budget), December 29, 1971

Dear Cap:

The purpose of this letter is to report the results of recent studies of several space shuttle options, and to recommend a course of action to be taken in the FY 1973 budget.

SUMMARY

We have concluded that the full capability 15 x 60´–65,000# [pound] payload shuttle still represents a "best buy," and in ordinary times should be developed. However, in recognition of the extremely severe near-term budgetary problems, we are recommending a somewhat smaller vehicle—one with a 14 x 45´–45,000# payload capability, at a somewhat reduced overall cost.

This is the smallest vehicle that we can still consider to be useful for manned flight as well as a variety of unmanned payloads. However, it will not accommodate many DOD payloads and some planetary payloads. Also, it will not accommodate a space tug together with a payload, and will therefore not provide an effective capability to return payloads or propulsive stages from high "synchronous" orbits, where most applications payloads are placed.

***

DECISION TO PROCEED

The various shuttle studies have progressed to the point where a decision to proceed with full shuttle development should now be made. Further delays would not produce significant new results. The orbiter is fully defined. Although a question of solid versus liquid boosters remains open, the range of variables involved in the booster decision is not large, and a decision can be made at an early date. No substantial cost savings can be realized by further studies. (All of the most recent cost refinements for a given payload size have been less than the overall cost uncertainties inherent in a large R&D undertaking.)

On the other hand, additional delays would have many unsettling effects. In the aerospace industry, the existing shuttle teams will soon be dissipated, unless fully funded by the government. Last year's strong Congressional support for the shuttle may be lost this year if the Administration cannot present equally strong support. And within NASA, many of the best people will be lost, with a resulting loss in overall morale.

In other words, there is a great deal to be gained, and nothing to be lost, by making a decision to proceed now.

***

RESULTS OF STUDIES

CASE	1	2	2A	3	4
PAYLOAD BAY (FT.)	10 x 30	12 x 40	14 x 45	14 x 50	15 x 60
PAYLOAD WEIGHT (LBS.)	30,000	30,000	45,000	65,000	65,000
DEVELOPMENT COST (BILLIONS)	4.7	4.9	5.0	5.2	5.5
OPERATING COST ($MILLION/FLT.)	6.6	7.0	7.5	7.6	7.7
PAYLOAD COSTS ($/POUND)	220	223	167	115	118

The debate over which shuttle configuration to approve continued over the New Year's weekend. On January 3, 1972, NASA was surprised to learn that it had received OMB approval to develop the full-capability shuttle with a fifteen-foot-by-sixty-foot payload bay, rather than the smaller system recommended in Fletcher's December 29 letter.

Low and Fletcher flew to California to meet on January 5 with President Nixon, who was at the Western White House in San Clemente, prior to a presidential announcement of the shuttle project. This memorandum records George Low's version of the meeting with the president. After the meeting, the White House announced approval of the shuttle to the press, and Fletcher and Low answered questions about the project. With the approval of the space shuttle, the post-Apollo human spaceflight program for the foreseeable future would be limited to low-Earth-orbit missions.

George M. Low (Deputy Administrator, NASA), Memorandum for the Record, "Meeting with the President on January 5, 1972," January 12, 1972

Jim Fletcher and I met with the President and John Ehrlichman for approximately 40 minutes to discuss the space shuttle. During the course of the discussion, the President either made or agreed with the following points:

1. **The Space Shuttle.** The President stated that we should stress civilian applications but not to the exclusion of military applications. We should not hesitate to mention the military applications as well. He was interested in the possibility of routine operations and quick reaction times, particularly as these would apply to problems of natural disasters, such as earthquakes or floods. When Dr. Fletcher mentioned a future possibility of collecting solar power in orbit and beaming it down to earth, the President indicated that these kinds of things tend to happen much more quickly than we now expect and that we should not hesitate to talk about them now. He was also interested in the nuclear waste disposal possibilities. The President liked the fact that ordinary people would be able to fly in the shuttle, and that the only requirement for a flight would be that there is a mission to be performed. He also reiterated his concern for preserving the skills of the people in the aerospace industry.

   In summary, the President said that even though we now know of many things that the shuttle will be able to do, we should realize that it will open up entirely new fields when we actually have the capability that the shuttle will provide. The President wanted to know if we thought the shuttle was a good investment and, upon receiving our affirmative reply, requested that we stress the fact that the shuttle is not a "$7 billion toy," that it is indeed useful, and that it is a good investment in that it will cut operations costs by a factor of 10. But he indicated that even if it were not a good investment, we would have to do it anyway, because space flight is here to stay. Men are flying in space now and will continue to fly in space, and we'd best be part of it.

2. **International Cooperation.** The President said that he is most interested in making the space program a truly international program and that he had previously expressed that interest. He wanted us to stress international cooperation and participation for *all* nations. He said that he was disappointed that we had been unable to fly foreign astronauts on Apollo, but understood the reasons for our inability to do so. He understood that foreign astronauts of all nations could fly in the shuttle and appeared to be particularly

interested in Eastern European participation in the flight program. However, in connection with international cooperation, he is not only interested in flying foreign astronauts, but also in other types of meaningful participation, both in experiments and even in space hardware development.

3. <u>USSR Cooperation.</u> The President was interested in our joint activities with the USSR in connection with the probes now in orbit around Mars. We also described to him the real possibility of conducting a joint docking experiment in the 1975 time period. The prospect of having Americans and Russians meet in space in this time period appeared to have great appeal to the President. He indicated that this should be considered as a possible item for early policy level discussions with the USSR. The President asked John Ehrlichman to mention both the international aspects of the shuttle and the USSR docking possibilities to Henry Kissinger.

While NASA focused on developing the space shuttle during the 1970s, there were two spaceflight efforts using excess Apollo hardware. One was Skylab, an experimental space station using the converted upper stage of a Saturn V booster. Skylab was launched on May 14, 1973, and was occupied by three three-person crews between May 25, 1973, and February 8, 1974. The crews stayed aboard Skylab for twenty-eight days, then fifty-six days, and eighty-four days. Skylab was not designed for resupply, and made an uncontrolled reentry into the Earth's atmosphere on July 11, 1979. Pieces of the vehicle fell to Earth in Western Australia.

The other flight project was the docking between Soviet and U.S. spacecraft discussed with President Nixon in his January 5, 1972, meeting with Fletcher and Low. NASA had been discussing such a project with its Soviet counterparts since 1970. With the president's expressed interest, an agreement on a docking mission was soon crafted for Nixon and his Soviet counterpart, Premier Alexei Kosygin, to sign at their May 1972 summit meeting. This agreement initiated what became known as the Apollo-Soyuz Test Project. On July 17, 1975, astronaut Tom Stafford and cosmonaut Alexey Leonov shook hands as their docked spacecraft met in orbit.

Following the success of the Apollo-Soyuz Test Project, NASA and the Soviet Academy of Sciences on May 11, 1977, signed a follow-on agreement that anticipated continuing cooperation in human spaceflight. However, President Jimmy Carter later decided on the basis of the Soviet human rights record and Soviet intervention in Afghanistan not to implement this agreement. The activities it presciently anticipated in 1977, the docking of the U.S. space shuttle with a Soviet space station and joint planning of an international space station, did not take place until the 1990s, after the fall of the Soviet Union.

During the brief presidency of Gerald Ford, there were no major human spaceflight issues raised. Developing the space shuttle had given NASA a major focus for its activities during the 1970s, and Ford simply approved the plans set out by the Nixon administration. The 1975 Apollo-Soyuz Test Project mission did take place while Ford was in the White House.

Technical problems with shuttle development emerged, delaying the shuttle's first flight by several years past the original target date of 1978. Jimmy Carter was not a space enthusiast; he was unconvinced, in particular, of the value of human spaceflight. He considered canceling the shuttle, but was persuaded that the vehicle, as the planned sole U.S. means of access to space after it began operations, was critical to national security, and thus its development had to continue.

In May 1978, Carter ordered a comprehensive review of the government's civilian space efforts. NASA and its allies, recognizing that shuttle development was close to completion, were beginning to seek approval of a space station to follow the shuttle. The Carter review took a negative position on such a possibility, saying that future space efforts would not center "around a single, massive engineering feat." There would be no new human spaceflight initiatives during the Carter administration.

Ronald Reagan was elected president in November 1980. He was told by his NASA transition team that the U.S. space program was at a "crossroads." With the high inflation that characterized the U.S. economy during the 1970s, NASA's budget had suffered a significant decline in purchasing power. The team's report provided a detailed set of recommendations and actions for the incoming administration, suggesting that presidential engagement was needed to restore U.S. space leadership. The team *recommended that "the purpose and direction of the US space effort be defined, and that a commitment to a viable space program be articulated by the President" at an early opportunity.*

George M. Low (Team Leader), "Report of the Transition Team, National Aeronautics and Space Administration," December 19, 1980

A. Overview

In 1958 the people of the United States set out to lead the world in space. By 1970 they had achieved their goal. Men walked on the moon, scientific satellites opened new windows to the universe, and communications satellites and new technologies brought economic return. With these came new knowledge and ideas, a sense of pride, and national prestige. In 1980, by contrast, United States leadership and preeminence are seriously threatened and measurably eroded. The Soviet Union has established an essentially permanent manned presence in space, and is using this presence to meet economic, military, and foreign policy goals. Japan is broadcasting directly from space to individual homes and business, and France is moving ahead of the United States in preparing to reap the economic benefits of satellite resource observation. Ironically, U.S. commercial enterprises are turning to France to launch their satellites. In space science, the United States has decided to forego the rare opportunity to visit Halley's comet in 1986, yet the Soviet Union, the European Space Agency, and Japan are all planning such a venture. Technically, it is within our means to reestablish U.S. preeminence in space. The civil space program and the National Aeronautics and Space Administration offer a number of options to carry out the purpose and direction of U.S. aeronautics and space activities. These options are examined in this report in full recognition of the need for fiscal restraint in the immediate future.

***

D. The Space Program and U.S. Policy

In recent years the United States has lost its competitive edge in the world, militarily, commercially, and economically, and our competition with the Soviet Union has taken on a new dimension. The Soviet Union recognizes that science and technology are major factors in that competition. The nation that is strong in science and technology has the foundation to be strong in all other areas and will be perceived as a world leader. Aeronautics and space can be major factors in our technological strength. They demand the very best in engineering, because the consequences of mistakes are great: the crash of an aircraft, or the complete failure of a spaceship. A viable aviation industry and a strong space program are important visible elements in our international competition. Beyond these fundamental points, the United States civil space program, unlike many other government programs and agencies, has significant actual and potential impact on U.S. policy. Although some elements of the program have been so utilized, their potential in U.S. policy remains largely unrecognized and unrealized. The major factors are as follows:

1. National Pride and Prestige

National pride is how we view ourselves. Without a national sense of purpose and identity, national pride ebbs and flows in accordance with short-term events. The Iranian hostage situation and the abortive rescue mission have done harm to our national pride quite out of proportion to our true abilities as a nation. On the other hand, the recent Voyager visit to Saturn, reported by an enthusiastic press, made a significant contribution to our sense of self-worth. The space program has characteristics of American historic self-image: a sense of purpose; a pioneering spirit of exploration, discovery, and adventure; a challenge of frontiers and goals; a recognition of individual contributions and team efforts; and a firm sense of innovation and leadership. National prestige is how others view us, the global perception of this country's intellectual, scientific, technological, and organizational capabilities. In recent history, the space program has been the unique positive factor in this regard. The Apollo exploration of the Moon restored our image in the post-Sputnik years, and the Voyager exploration of Saturn was a bright spot in an otherwise gloomy period of dwindling world recognition. With space programs we are a nation of the present and the future, while in the eyes of the world we become outward and forward looking.

2. Economics and Space Technology

A vigorous space program has provided many technological challenges to our nation. Efforts such as Apollo, Voyager, and the Space Shuttle have involved challenges and risks far more significant than those of short-term technological needs. Meeting these challenges has resulted in a "technological push" to American industry, fostering significant innovation in a wide range of high technology fields such as electronics, computers, science, aviation, communications and biomedicine. The return on the space investment is higher productivity, and greater competitiveness in the world market. The space program also returns direct dividends, as in the field of satellite communications. The potential economic returns from satellite exploration for earth's resources are great.

3. Scientific Knowledge and Inspiration for the Nation's Youth

U.S. leadership in the scientific exploration of space has provided new knowledge about the earth and the universe, thus forming the basis for applied research and development—a significant factor in our society and economy. The exploration of space has provided an inspirational focus for large numbers of young people who have become students of engineering and science. At a time when there is a shortage of technically trained people, when the U.S. productive vitality depends on the application of science, the space program could help attract young people into these fields.

4. Relation to U.S. Foreign Policy

Aspects of the civil space program can serve as instruments to develop and further U.S. foreign policy objectives. Not only can the space program contribute to how this country is viewed in the eyes of the world, but cooperative space activities, such as the U.S.-U.S.S.R. Apollo-Soyuz mission and European Space Agency payloads on the Space Shuttle, are important to other countries. Technology associated with the space program has resulted in strong economic and technological interaction with developed countries, as well as in important aid to underdeveloped countries, particularly in the areas of communications and resource exploration.

E. Observations

At the end of 1980 the U.S. civil space program stands at a cross-roads. The United States has invested in a great capability for space exploration and applications, a capability that provides benefits in national pride and prestige, in science and technology, in the inspiration of young people, in foreign policy, and in economic gain. Now this capability is waning. NASA and the space program are without clear purpose or direction.

VI. SUMMARY OF RECOMMENDATIOMS

NASA represents an important investment by the United States in aeronautics and space. The agency's programs have provided, and continue to offer, benefits in science and technology, in national pride and prestige, in foreign policy, and in economic gain. However, in recent years the agency has been underfunded, without purpose or direction. The new administration finds NASA at a crossroads, with possible moves toward either retrenchment or growth. The transition team has examined ten major areas and various options for dealing with them. For each issue, the team has made recommendations as follows:

A. *Presidential statement of purpose of the U.S. civil space program* . . . It is recommended:

1. That the President recognize the importance of the U.S. space program at an early date (e.g., the inaugural address) without yet making a commitment.

2. That the purpose and direction of the U.S. space effort be defined, and that a commitment to a viable space program be articulated by the President at a timely opportunity, such as the first flight of the Shuttle in the spring of 1981. (N.B. A viable space program could be smaller than, equal to, or larger than the present one, but it must have purpose and direction.)

***

This advice was not taken. Given his overriding concern with reducing the federal budget and other pressing issues on his agenda, Ronald Reagan did not provide during his eight years in the White House a clear articulation of the "purpose and direction of the U.S. space effort," although he did approve a major space initiative, the space station, and gave greatly increased attention to private sector space activities.

From the start of space shuttle development, the policy assumption had been that once it entered operation, the shuttle would be used to launch all government payloads, not only for NASA but also for other civilian agencies and the national security community. President Jimmy Carter had formalized this assumption in his 1978 national space policy directive. As Ronald Reagan became president in January 1981 and as the first launch of the shuttle was scheduled for April of that year, the premise of dependence on the shuttle was once again examined. Some in the military and intelligence sectors were concerned about sole reliance for access to space on a single launch system, but they were not successful in changing the existing policy. In November 1981, the Reagan administration committed itself to continuing the policy of making the shuttle the "primary" U.S. launch system. The Space Transportation System (STS) was the designation used for the total space shuttle system, including any upper stages needed to move a payload to its final orbit.

The first launch of the space shuttle took place on April 12, 1981. Three more test missions followed, and as the fourth shuttle mission landed on July 4, 1982, with the president and his wife in attendance, the shuttle was declared "operational." This meant that NASA could now solicit commercial and foreign customers to use the shuttle, on a fee-paying basis, to launch their payloads aboard the vehicle. By 1982, NASA and the shuttle were competing with the European expendable launch vehicle Ariane for such contracts. The competition took on a chauvinistic tone.

To help in this competition and in marketing the shuttle overall, NASA published a glossy brochure titled "We Deliver." The sales-oriented language in the brochure marked a dramatic change from NASA's heritage as a research and development organization.

NASA, "We Deliver," 1983

YOU CAN'T GET A BETTER PRICE

The price for launching a payload on the Shuttle is based on the share of the lifting weight and cargo bay length required by your payload. Pricing according to this continuous curve formula assures that you will be charged only for your requirements, while retaining the significant growth capability inherent in the Shuttle in the event those requirements change during the period you are developing your payload. The Shuttle price and pricing flexibility cannot be matched by any other launch system.

In evaluating the total cost associated with a launch, two other important factors must be considered, both involving insurance. One is the effect that NASA's launch record and experience has had on insurance rates. It has been demonstrated that this record commands for the Shuttle the lowest insurance rates in the free world. A difference of a few percentage points in the insurance rates for a satellite or other payload program can easily mean millions of dollars saved in the insurance purchased for your launch.

The other insurance consideration involves the charges associated with postponing a payload launch on the Shuttle. We have a commitment to all of our customers to launch on time, and we recognize the importance of their cash-flow demands. Therefore, as an incentive for customers to make all reasonable effort to have their payloads delivered for launch on the agreed schedule, we have established significant postponement fees. On the other hand, we appreciate the cost risk associated with postponements. Here again, the insurance industry has demonstrated its support of the Shuttle by agreeing to provide insurance, at low premium rates, to cover the postponement fees.

Considering all cost factors associated with launching your satellite or other payload into space, you can't get a better price or more for your money than the Space Shuttle.

The Reagan administration in 1981 chose James Beggs as its NASA administrator. From the start of his time in that position, Beggs and his deputy Hans Mark had made getting presidential approval for a major new engineering project, a permanently crewed orbital space station, one of their highest priorities. This was a legacy of the 1970 decision to develop the space shuttle and space station one after the other rather than simultaneously. Approving the space station was opposed by the national security community and the White House budget and science offices, and it took until the end of 1983, and only after a yearlong period of interagency study, to bring the proposition before the president. On December 1, 1983, in the White House Cabinet Room, James Beggs made a presentation to the president, saying, "Today, the Space Shuttle makes us the leading Nation in space. Tomorrow, America's preeminence in space can be achieved through a space station, manned and in permanent orbit around the earth. Shuttle is routine transportation to Earth orbit. What's needed now, what was originally envisioned, is a place to Shuttle to. . . ." At the end of his presentation, Beggs made an impassioned plea: "Now, today, here in this room, we must look forward to the next twenty-five years. The time to start a space station is now: Shuttle development is over; technology is at hand; requirements have been analyzed; industry is ready; lead times are long; the stakes are enormous: leadership in space for the next twenty-five years."

After Beggs concluded his presentation, the president asked for comments from others in the room. This was Ronald Reagan's preferred approach to decision making. He wanted to hear the views of his top advisers before making a decision. Most of the comments around the table opposed space station approval. As was his practice, Reagan did not announce his decision at the conclusion of the meeting.

One of the high-level officials attending the Cabinet Room meeting was John McMahon, deputy director of the Central Intelligence Agency. He prepared a summary of the discussion following Beggs's presentation.

John McMahon (Deputy Director of the Central Intelligence Agency), Memorandum for the Record, "Summary of Cabinet Council on Trade and Commerce Meeting, 1 December 1983," December 5, 1983

1. The meeting was chaired by Secretary [of Commerce Malcolm] Baldrige. The President, Vice President, et al, were in attendance. Gil Rye [National Security Director for space] introduced the subject and summed up the results of the SIG (Space) meetings. However, he didn't indicate the voting of the SIG which parenthetically was overwhelmingly against the manned space station (MSS). He did indicate that it would cost about $8 billion over the next 7 years, requiring $225 million in 1984. Another option was to extend the shuttle from 7 to 21 days at a cost of $3.1 billion with $190 million needed in 1984.

2. Beggs then spoke, gave assurance that the number 1 priority for NASA is to make the shuttle fully operational. He said that some work has to be done and that they would certainly put their attention to it. This was obviously said to mute any criticism that might come from Defense that NASA ought to make right the programs it has now before it launches off into something new. Beggs then said that 90¢ out of every dollar at NASA winds up in the private sector or academia. He then spoke flatteringly of the manned space station, said that they hoped to have an IOC by 1991 or 1992. He then launched into an argument on why NASA must reach out and that the United States should capitalize on what NASA has done so far and move out into the space station. He noted in 1903, two Americans flew the first aircraft. However, nothing was really done with it for another eleven years and when World War I came the U.S. had to rely on European planes to fly. Following that lesson NACA was formed which did experiments in aerodynamics and when World War II came we were well prepared to fly our own aircraft. He noted that in 1926 Goddard launched the first rocket up in Massachusetts, that we as a nation failed to capitalize on that technology but the Germans did and in World War II they flew the rockets and we didn't.

3. The President interrupted Beggs by asking him what about a moon base. Beggs replied that the MSS is an interim station to the goal of a moon base. First comes the MSS, then a moon station manned by robotics and then finally a man-based moon station.

4. George Keyworth [Reagan science adviser] commented that the MSS is not new and he feels that we shouldn't emphasize a simple way station but rather go for a bold step and optimize for a goal which would mean something in space, advocating a moon space station. He readily admitted, however, that he didn't know what we would do on the moon or how we would get there but suggested over the next six months we study it to determine those answers.

5. Baldrige then commented to the President that the options we considered were too narrow and probably we ought to undertake a study into the commercial and civil use of space—in other words do a survey to that end.

6. [Paul] Thayer [Deputy Director of Defense] then spoke of the impact that the MSS would have on diverting NASA's attention away from their problems with the shuttle and optimizing that for the purpose for which it was designed. He also felt that the MSS would not be an $8 billion program but rather a $20 billion one. He noted that there were no military requests for the MSS—that it has no value for the Pentagon. He also noted that the National Academy of Science and the Chamber of Commerce have come out against the MSS. If we had the money maybe we could take the chance and do a thing like the MSS but there is a shortage of funds to do essential items and we can't visualize the MSS in any way recouping its investment. He said because of his own uncertainties on the future demands for dollars he proposed not scheduling the MSS now but deferring it for six months to a year so that we would have a better picture of resources needed.

7. [David] Stockman [OMB director] then spoke, said that the MSS was excessive in costs when resources are scarce. He felt that it would cost at least $10 to $15 billion and that amount would push the budget to fracture. He noted that the $3.5 trillion program puts us out perilously close to

fiscal bankruptcy. He noted for the President that in FY 1983 we had a $208 billion deficit. In 1984, Congress didn't cut one dime and we will experience another $200 billion deficit. Fiscal 1985 will be driven by an election year and we will undoubtedly have a $200 billion deficit then. Thus, in just three fiscal years we will have added $600 billion to the national debt. He said that you, Mr. President, have got to get ahold of Congress and force them to do what they ought to do. Then maybe in FY 1986 we can cut the deficit to $156 billion and by 1987 to $150 billion. But what this means is that the mid-80's will add $1 trillion to our national debt.

8. The President said that he didn't plan to make a decision today but just wanted to hear evryone [sic] out. Baldrige commented that we haven't heard yet from the private sector and that there is some talk that several companies may form a consortium to build the MSS and lease it back to the U.S. This generated a few smiles and smirks around the table.

9. Bill Brock, U.S. Trade Representative, then spoke. He claimed that we really haven't spent any dollars in the past 20 years because every dollar spent in space has been returned to us 20 times over in economic growth and flow of dollars to the treasury. McNamar [Deputy Secretary of the Treasury] interrupted him by saying that the Treasury has yet to see its first dollar from space and if it's true that we can get 20 to 1 for an investment like the MSS then he and Donald Regan are going to leave the Treasury and set up a company to do that. The Attorney General interrupted by saying that if Ferdinand and Isabella had the same attitude the U.S. would not have been discovered. The President interrupted by saying that Isabella had jewelry to sell and he doesn't. Thayer commented, "Where is Spain now?" Brock then gave an emotional talk on how dollars spent on welfare are wasted, we never see them and even noted how the seed corn that we gave the Indians which would have increased their harvest tenfold was eaten instead of planted.

10. Craig Fuller [White House director of cabinet affairs] then commented about how he must represent U.S. commercial interests with whom he has been in touch and he feels that

if the President makes this commitment that U.S. industry is with him and that he would even have our European partners cheering us on.

11. Whereupon the meeting ended and we all withdrew.

On January 25, 1984, Reagan announced his approval of a space station. He had actually approved the NASA proposal four days after Beggs's December 1 presentation, but held off announcing his approval until his State of the Union address. In the speech, he also invited U.S. friends and allies to participate in the space station, tying the human spaceflight ambitions of those countries to the fate of the U.S. space station effort.

Ronald Reagan, "State of the Union Address," January 25, 1984

***

It's time to move forward again, time for America to take freedom's next step. Let us unite tonight behind four great goals to keep America free, secure, and at peace in the eighties together.

***

Our second great goal is to build on America's pioneer spirit . . .

Nowhere is this more important than our next frontier: space. Nowhere do we so effectively demonstrate our technological leadership and ability to make life better on Earth. The Space Age is barely a quarter of a century old. But already we've pushed civilization forward with our advances in science and technology. Opportunities and jobs will multiply as we cross new thresholds of knowledge and reach deeper into the unknown.

Our progress in space—taking giant steps for all mankind—is a tribute to American teamwork and excellence. Our finest minds in government, industry, and academia have all pulled together. And we can be proud to say: We are first; we are the best; and we are so because we're free.

America has always been greatest when we dared to be great. We can reach for greatness again. We can follow our dreams to distant stars, living and working in space for peaceful, economic, and scientific gain. Tonight, I am directing NASA to develop a permanently manned space station and to do it within a decade.

A space station will permit quantum leaps in our research in science, communications, in metals, and in lifesaving medicines which could be manufactured only in space. We want our friends to help us meet these challenges and share in their benefits. NASA will invite other countries to participate so we can strengthen peace, build prosperity, and expand freedom for all who share our goals.

***

On January 28, 1986, due to a failure of the seals in a joint on the shuttle's right solid fuel rocket motor, the shuttle orbiter Challenger's *external fuel tank was breached, allowing highly flammable hydrogen and oxygen to escape. The liquids ignited seventy-two seconds after launch. The resulting conflagration caused the shuttle to break apart; its seven-person crew perished as the orbiter's crew compartment plunged 65,000 feet into the Atlantic Ocean.*

That evening, President Ronald Reagan addressed the nation.

Ronald Reagan, "Address to the Nation on the Explosion of the Space Shuttle *Challenger*," January 28, 1986

Ladies and gentlemen, I'd planned to speak to you tonight to report on the state of the Union, but the events of earlier today have led me to change those plans. Today is a day for mourning and remembering. Nancy and I are pained to the core by the tragedy of the shuttle Challenger. We know we share this pain with all of the people of our country. This is truly a national loss.

Nineteen years ago, almost to the day, we lost three astronauts in a terrible accident on the ground. But we've never lost an astronaut in flight; we've never had a tragedy like this. And perhaps we've forgotten the courage it took for the crew of the shuttle. But they, the Challenger Seven, were aware of the dangers, but overcame them and did their jobs brilliantly. We mourn seven heroes: Michael Smith, Dick Scobee, Judith Resnik, Ronald McNair, Ellison Onizuka, Gregory Jarvis, and Christa McAuliffe. We mourn their loss as a nation together.

For the families of the seven, we cannot bear, as you do, the full impact of this tragedy. But we feel the loss, and we're thinking about you so very much. Your loved ones were daring and brave, and they had that special grace, that special spirit that says, "Give me a challenge, and I'll meet it with joy." They had a hunger to explore the universe and discover its truths. They wished to serve, and they did. They served all of us. We've grown used to wonders in this century. It's hard to dazzle us. But for 25 years the United States space program has been doing just that. We've grown used to the idea of space, and perhaps we forget that we've only just begun. We're still pioneers. They, the members of the Challenger crew, were pioneers.

And I want to say something to the schoolchildren of America who were watching the live coverage of the shuttle's takeoff. I know it is hard to understand, but sometimes painful things like this happen. It's all part of the process of exploration and discovery. It's all part of taking a chance and expanding man's horizons. The future doesn't belong to the fainthearted; it belongs to the brave. The Challenger crew was pulling us into the future, and we'll continue to follow them.

I've always had great faith in and respect for our space program, and what happened today does nothing to diminish it. We don't hide our space program. We don't keep secrets and cover things up. We do it all up front and in public. That's the way freedom is, and we wouldn't change it for a minute. We'll continue our quest in space. There will be more shuttle flights and more shuttle crews and, yes, more volunteers, more civilians, more teachers in space. Nothing ends here; our hopes and our journeys continue. I want to add that I wish I could talk to every man and woman who works for NASA or who worked on this mission and tell them: "Your dedication and professionalism have moved and impressed us for decades. And we know of your anguish. We share it."

There's a coincidence today. On this day 390 years ago, the great explorer Sir Francis Drake died aboard ship off the coast of Panama. In his lifetime the great frontiers were the oceans, and an historian later said, "He lived by the sea, died on it, and was buried in it." Well, today we can say of the Challenger crew: Their dedication was, like Drake's, complete.

The crew of the space shuttle Challenger honored us by the manner in which they lived their lives. We will never forget them, nor the last time we saw them, this morning, as they prepared for their journey and waved goodbye and "slipped the surly bonds of earth" to "touch the face of God."

In the aftermath of the January 1967 Apollo 1 fire, NASA had persuaded the White House to allow it to manage the accident investigation. In January 1986 the White House decided that it should appoint a board independent of NASA to investigate the Challenger *accident. The makeup of the board was announced on February 3; the group would be chaired by former secretary of state William Rogers, and soon became known as the Rogers Commission. The commission submitted its 261-page final report on June 6, 1986.*

"Report to the President by the Presidential Commission on the Space Shuttle *Challenger* Accident," June 6, 1986

PREFACE

***

The President, who was moved and troubled by this accident in a very personal way, appointed an independent Commission made up of persons not connected with the mission to investigate it. The mandate of the Commission was to:

1. Review the circumstances surrounding the accident to establish the probable cause or causes of the accident; and
2. Develop recommendations for corrective or other action based upon the Commission's findings and determinations.

***

However, the Commission did not construe its mandate to require a detailed investigation of all aspects of the Space Shuttle program; to review budgetary matters; or to interfere with or supersede Congress in any way in the performance of its duties. Rather, the Commission focused its attention on the safety aspects of future flights based on the lessons learned from the investigation with the objective being to return to safe flight.

***

CHAPTER IV: THE CAUSE OF THE ACCIDENT

***

Findings

1. A combustion gas leak through the right Solid Rocket Motor aft field joint initiated at or shortly after ignition eventually weakened and/or penetrated the External Tank initiating vehicle structural breakup and loss of the Space Shuttle Challenger during STS Mission 51-L.

2. The evidence shows that no other STS 51-L Shuttle element or the payload contributed to the causes of the right Solid Rocket Motor aft field joint combustion gas leak. Sabotage was not a factor.

***

5. Launch site records show that the right Solid Rocket Motor segments were assembled using approved procedures. However, significant out-of-round conditions existed between the two segments joined at the right Solid Rocket Motor aft field joint (the joint that failed).

    a. While the assembly conditions had the potential of generating debris or damage that could cause O-ring seal failure, these were not considered factors in this accident.

    b. The diameters of the two Solid Rocket Motor segments had grown as a result of prior use.

    c. The growth resulted in a condition at time of launch wherein the maximum gap between the tang and clevis in the region of the joint's O-rings was no more than .008 inches and the average gap would have been .004 inches.

    d. With a tang-to-clevis gap of .004 inches, the O-ring in the joint would be compressed to the extent that it pressed against all three walls of the O-ring retaining channel.

    e. The lack of roundness of the segments was such that the smallest tang-to-clevis clearance occurred at the initiation of the assembly operation at positions of 120 degrees and 300 degrees around the circumference of the aft field joint. It is uncertain if this tight condition and the resultant greater compression of the O-rings at these points persisted to the time of launch.

6. The ambient temperature at time of launch was 36 degrees Fahrenheit, or 15 degrees lower than the next coldest previous launch.

   a. The temperature at the 300 degree position on the right aft field joint circumference was estimated to be 28 degrees +/–5 degrees Fahrenheit. This was the coldest point on the joint.

***

9. O-ring resiliency is directly related to its temperature.

   a. A warm O-ring that has been compressed will return to its original shape much quicker than will a cold O-ring when compression is relieved. Thus, a warm O-ring will follow the opening of the tang-to-clevis gap. A cold O-ring may not.

   b. A compressed O-ring at 75 degrees Fahrenheit is five times more responsive in returning to its uncompressed shape than a cold O-ring at 30 degrees Fahrenheit.

   c. As a result it is probable that the O-rings in the right solid booster aft field joint were not following the opening of the gap between the tang and clevis at time of ignition.

***

12. Of 21 launches with ambient temperatures of 61 degrees Fahrenheit or greater, only four showed signs of O-ring thermal distress; i.e., erosion or blow-by and soot. Each of the launches below 61. degrees Fahrenheit resulted in one or more O-rings showing signs of thermal distress.

   a. Of these improper joint sealing actions, one-half occurred in the aft field joints, 20 percent in the center field joints, and 30 percent in the upper field joints. The division between left and right Solid Rocket Boosters was roughly equal. Each instance of thermal

O-ring distress was accompanied by a leak path in the insulating putty. The leak path connects the rocket's combustion chamber with the O-ring region of the tang and clevis. Joints that actuated without incident may also have had these leak paths.

***

14. A series of puffs of smoke were observed emanating from the 51-L aft field joint area of the right Solid Rocket Booster between 0.678 and 2.500 seconds after ignition of the Shuttle Solid Rocket Motors.

***

15. This smoke from the aft field joint at Shuttle lift off was the first sign of the failure of the Solid Rocket Booster O-ring seals on STS 51-L.
16. The leak was again clearly evident as a flame at approximately 58 seconds into the flight.

***

Conclusion

In view of the findings, the Commission concluded that the cause of the Challenger accident was the failure of the pressure seal in the aft field joint of the right Solid Rocket Motor. The failure was due to a faulty design unacceptably sensitive to a number of factors. These factors were the effects of temperature, physical dimensions, the character of materials, the effects of reusability, processing, and the reaction of the joint to dynamic loading.

With its role in operating the space shuttle and approval of space station development, NASA had a full plate of activities to occupy it in coming years. But the space station, like the space shuttle, was a means to an end. Reagan did not spell out a long-range goal that the space station was to serve. This was similar to the situation in 1972, when Richard Nixon had approved space shuttle development rather than set an ambitious post-Apollo goal. This was evident to some in the White House, Congress, and the external space community, who had argued during the space station debate that the president should set out such long-range objectives.

During this debate, presidential science adviser George A. Keyworth called for a White House symposium to discuss future space goals, and Congress later in 1984 passed legislation calling for President Reagan to establish a National Commission on Space to set out a long-range vision. That fifteen-member commission was to look twenty years into the future, but its chair, former NASA administrator Thomas O. Paine, instead mandated that the group look fifty years ahead.

The commission's report was not presented to President Reagan until mid-1986. By that time, the country was struggling with how best to recover from the space shuttle Challenger *accident, and there was little appetite for broad, long-term visions for the space program.*

National Commission on Space, "Pioneering the Space Frontier: An Exciting Vision of Our Next Fifty Years in Space," 1986

A PIONEERING MISSION FOR 21ST-CENTURY AMERICA

To lead the exploration and development of the space frontier, advancing science, technology, and enterprise, and building institutions and systems that make accessible vast new resources and support human settlements beyond Earth orbit, from the highlands of the Moon to the plains of Mars

RATIONALE FOR EXPLORING AND SETTLING THE SOLAR SYSTEM

Our Vision: The Solar System as the Home of Humanity

The Solar System is our extended home. Five centuries after Columbus opened access to "The New World" we can initiate the settlement of worlds beyond our planet of birth. The promise of virgin lands and the opportunity to live in freedom brought our ancestors to the shores of North America. Now space technology has freed humankind to move outward from Earth as a species destined to expand to other worlds.

Our Purpose: Free Societies on New Worlds

The settlement of North America and other continents was a prelude to humanity's greater challenge: the space frontier. As we develop new lands of opportunity for ourselves and our descendants, we must carry with us the guarantees expressed in our Bill of Rights: to think, communicate, and live in freedom. We must stimulate individual initiative and free enterprise in space.

Our Ambition: Opening New Resources to Benefit Humanity

Historically, wealth has been created when the power of the human intellect combined abundant energy with rich material resources. Now America can create new wealth on the space frontier to benefit the entire human community by combining the energy of the Sun with materials left in space during the formation of the Solar System.

Our Method: Efficiency and Systematic Progression

In undertaking this great venture we must plan logically and build wisely. Each new step must be justified on its own merits and make possible additional steps. American investments on the space frontier should be sustained at a small but steady fraction of our national budget.

Our Hope: Increased World Cooperation

In his essay *Common Sense*, published in January of 1776, Tom Paine said of American independence, "'Tis not the affair of a City, County, a Province, or a Kingdom; but of a Continent . . . 'Tis not the concern of a day, a year, or an age; posterity are virtually involved in the contest, and will be more or less affected even to the end of time, by the proceedings now." Exploring the Universe is neither one nation's issue, nor relevant only to our time. Accordingly, America must work with other nations in a manner consistent with our Constitution, national security, and international agreements.

Our Aspiration: American Leadership on the Space Frontier

With America's pioneer heritage, technological preeminence, and economic strength, it is fitting that we should lead the people of this planet into space. Our leadership role should challenge the vision, talents, and energies of young and old alike, and inspire other nations to contribute their best talents to expand humanity's future.

Our Need: Balance and Common Sense

Settling North America required the sustained efforts of laborers and farmers, merchants and ministers, artisans and adventurers, scientists and seafarers. In the same way, our space program must combine with vigor and continuity the elements of scientific research, technological advance, the discovery and development of new resources in space, and the provision of essential institutions and systems to extend America's reach in science, industry, and the settlement of space.

Our Approach: The Critical Lead Role of Government

As formerly on the western frontier, now similarly on the space frontier, Government should support exploration and science, advance critical technologies, and provide the transportation systems and administration required to open broad access to new lands. The investment will again generate in value many times its cost to the benefit of all.

Our Resolve: To Go Forth "In Peace for All Mankind"

When the first Apollo astronauts stepped onto the Moon, they emplaced a plaque upon which were inscribed the words, "We came in peace for all mankind." As we move outward into the Solar System, we must remain true to our values as Americans: To go forward peacefully and to respect the integrity of planetary bodies and alien life forms, with equality of opportunity for all.

***

In the aftermath of the Challenger *accident, and after heated debate inside the Reagan administration, the central role of the space shuttle as the primary U.S. launch vehicle was significantly altered. The shuttle would no longer launch commercial satellites or compete for commercial and foreign launch contracts, and the national security community was authorized to procure expendable launch vehicles so that it could transition from its dependence on the shuttle. The shuttle, once it returned to flight on September 29, 1988, would launch almost exclusively NASA science payloads and eventually elements of the space station that Ronald Reagan had approved in 1984.*

Between its 1988 return to flight and its retirement from service in July 2011, the space shuttle was launched 109 times. Thirty-seven of those missions, beginning in 1998, were devoted to space station assembly and outfitting. One mission in 2003 ended in another catastrophic accident, killing the seven crewmembers. The shuttle launched a number of major space science payloads, including in April 1990 the Hubble Space Telescope. Soon after its launch, it was discovered that the telescope's primary mirror had been incorrectly ground, resulting in blurry images. After almost three years of planning and preparation, a high-profile Hubble servicing mission was launched in December 1993. That mission was successful in giving the telescope "contact lenses" so that its images became sharp and making Hubble what many have called the most important telescope in the history of astronomy. There were four more Hubble servicing missions between 1997 and 2009.

These daily mission status reports capture the complexity of the first repair mission.

Mission Control Center, "STS-61 Status Report #10," December 6, 1993

Mission Specialists Jeff Hoffman and Story Musgrove will put on their space work clothes for the second time about 9:47 p.m. CST tonight for a four-hour replacement of the Wide Field/Planetary Camera and a one-hour installation of two new magnetometers.

The seven-member crew of Endeavour was awakened at 6:02 p.m. Monday by flight controllers who played "Doctor My Eyes" by Jackson Browne.

Hoffman will step into a foot restraint on Endeavour's robot arm for the WF/PC swap, and Musgrave will be on a portable foot restraint anchored near the WF/PC opening on the Hubble Space Telescope. Both astronauts will be anchored on the end of the robot arm for the magnetometer installation.

Controllers at the Space Telescope Operations Control Center will begin powering down WF/PC I at 11:15 p.m., then begin reconfiguring their equipment to support the new WF/PC II.

On the aft flight deck, Mission Specialist Claude Nicollier will drive the robot arm, moving Hoffman into position to grasp WF/PC I. Musgrave will help stabilize the instrument as Hoffman slowly pulls WF/PC I out along its guide rails, pausing to allow Nicollier to reposition the arm. Before WF/PC I is completely removed, the trio will conduct a practice session to prepare for installation of WF/PC II. As Hoffman is placing the old camera in a temporary parking fixture in the payload bay, Musgrave will inspect the WF/PC orifice and begin preparing the new camera for removal from its transport container. The pair will then attach a transfer handle to the new camera and pull it out of the transport canister. Before Hoffman begins installing the new camera into the body of the telescope, Musgrave will remove a protective mirror cover. Then, the astronauts will carefully align the new camera on its guide rails and insert it into the telescope.

STOCC controllers will conduct an "aliveness" test on the new camera at 1:20 a.m. Tuesday, and begin functional tests about 4:40 a.m. A science data dump will recover images from the functional tests for processing by the WF/PC Instrument Development Team as early as 7:35 a.m. Results of those tests should be available within 30 minutes.

The telescope will be tilted forward on its work platform so that the robot arm can reach the top of the telescope, where the magnetometers are located. STOCC controllers will configure the first magnetometer for replacement about 1:40 a.m. After the astronauts install the first new unit, the STOCC will conduct functional tests about 3 a.m. After those tests are complete, the space walkers will install the second unit and the STOCC will conduct its functional tests about 3:20 a.m.

All of Endeavour's systems continue to perform well as the shuttle circles the Earth every 95 minutes in a 320 by 313 nautical mile orbit.

Mission Control Center, "STS-61 Status Report #11," December 7, 1993

The flawless installation of the Wide Field/Planetary Camera II early Tuesday morning highlighted the third back-to-back space-walk to service the Hubble Space Telescope.

"Ohhh, look at that baby, it's a beautiful spanking new Wif-pic," space walking astronaut Jeff Hoffman said as he pulled the replacement Wide Field/Planetary Camera II out of its storage locker about 11:24 p.m.

Following the removal and storage of the original WF/PC which will be returned to Earth for post-flight analysis, STS-61 space walking astronauts Story Musgrave and Hoffman installed the 620 pound camera about 12:05 a.m. CST Tuesday while in the Space Shuttle Endeavour's payload bay. The camera sits just below the telescope's midpoint. About 35 minutes after the installation was complete, ground controllers reported that the camera had pass the first electrical "aliveness" test, as it is called.

"I hope we have a lot of scientists eager to use this beautiful thing," Hoffman said after the installation. The space walking duo completed the installation of the camera in record time. Pre-flight predictions provided a four hour time slot for the detailed installation.

The original camera experienced focusing problems shortly after the telescope's deploy in April 1990. The problems were attributed to a manufacturing flaw in the telescope's 94-inch wide primary mirror. Blurred photographs were the result of the flaw. The new camera has four small precisely ground mirrors that should remove the blur by focusing the stray light of the tele-scope's primary reflector.

Hoffman and Musgrave also installed two new magnetometers during their six hour and 47 minutes spacewalk. The astronauts began the third spacewalk at 9:35 a.m. CST Monday, more than an hour earlier than planned pre-flight. The magnetometers, which are located at the top of the telescope, sense the magnetic field in three directions and are needed to keep the Hubble's momentum wheels operating with optimal efficiency.

At the end of today's third spacewalk, Musgrave had accumulated a total of 19 hours doing spacewalks and Hoffman had racked up a total of 17 hours and 51 minutes. Both astronauts have been on three separate spacewalks, two of which have occurred on STS-61. Musgrave was the first person in the shuttle program to conduct a spacewalk and he did so on STS-6. Hoffman's first spacewalk occurred on STS-51D.

Crew members will begin their sleep period at 9:57 a.m. CST and flight controllers will awaken them at 5:57 p.m. CST today. Musgrave will begin his seventh day in space with a television interview at 8:27 p.m. CST today. He will talk with Ted Koppel, the host of the ABC news program Nightline. The interview is expected to last about 15 minutes.

Following the interview, crew members will devote their attention to the fourth spacewalk scheduled for this mission. During tonight's spacewalk, astronauts Kathy Thornton and Tom Akers will replace the telescope's high-speed photometer with the Corrective Optics Space Telescope Axial Replacement. The COSTAR has 10 small mirrors that should properly focus light from the Hubble camera's primary reflectors. Thornton and Akers are scheduled to begin their spacewalk at 10:52 p.m. today.

All of Endeavour's systems continue to perform well as the shuttle circles the Earth every 95 minutes in a 320 by 313 nautical mile orbit.

As the shuttle was preparing to return to flight, the space station was having its own difficulties. By 1987 its estimated cost had increased by more than 75 percent, and congressional support for the project was wavering. Negotiations with potential international partners had veered near collapse. But after a careful review, President Reagan decided to proceed with the project. As one step in that direction, he approved NASA's proposal that the station be named "Freedom." Colin Powell was Reagan's national security adviser in 1988.

Colin L. Powell, Memorandum for the President, "Naming the Space Station," July 6, 1988

ISSUE

Whether you should approve a NASA recommendation to select and announce a name for the Space Station.

DISCUSSION

After almost three years of sometimes difficult negotiations, our close allies and friends in Western Europe, Japan, and Canada have agreed to join with us in a cooperative relationship that will substantially increase the capabilities of the U.S. Space Station. We anticipate that the intergovernmental agreements to seal that long-term partnership—the largest cooperative scientific and technological project ever undertaken—will be ready for signature during September 1988. This represents a major step in achieving the visionary goal you announced in your 1984 State of the Union Address: to develop and operate a permanently-manned civil space station.

As you know, the Space Station project is in some funding difficulty this year as a result of Congressional failure so far to appropriate the needed funds for NASA within the guidelines of the bipartisan budget agreement. Your continued support for the Space Station will be needed to secure the funding necessary to continue this important program at a funding level that will enable progress to be made toward the mid-1990s operational date.

NASA has recommended, and we agree, that your selection of a name for the Station will heighten public awareness and appreciation for the program during these difficult times. It also provides you with another opportunity to reaffirm your continued support for the program and explain its importance for America's future. Jim Fletcher has recommended that you approve the name "Freedom" for the Space Station. "Freedom" was the top choice of a selection team comprised of representatives from NASA and our international partners. It conveys the appropriate image for the West's space station, and complements nicely the Soviet name for their space station, Mir, which translates as "Peace."

The final Reagan administration space policy, issued in February 1988, set "the long-range goal of expanding human presence and activity beyond Earth orbit into the solar system." Reagan's successor, President George H. W. Bush, was well aware of the state of the U.S. human spaceflight program as he took office in January 1989. Bush was advised in his early months in office of widespread calls for setting a long-range goal for the U.S. space program.

Bush asked his vice president, Dan Quayle, to develop a proposal for achieving such a goal. After having been abolished by Richard Nixon in 1973, at the start of the Bush administration a National Space Council was reestablished at the White House level, with Quayle as its chair. It was in this role that the president asked Quayle to chair a rapid space program review. That review resulted in a bold proposal for resuming exploration beyond Earth orbit as the defining goal of U.S. spaceflight activity. President Bush accepted this proposal and, on July 20, 1989, the twentieth anniversary of the first lunar landing, on a hot July afternoon at the National Air and Space Museum, with the Apollo 11 crew by his side, announced what became known as the Space Exploration Initiative.

George H. W. Bush, "Remarks on the 20th Anniversary of the Apollo 11 Moon Landing," July 20, 1989

***

Space is the inescapable challenge to all the advanced nations of the Earth. And there's little question that, in the 21st century, humans will again leave their home planet for voyages of discovery and exploration. What was once improbable is now inevitable. The time has come to look beyond brief encounters. We must commit ourselves anew to a sustained program of manned exploration of the solar system and, yes, the permanent settlement of space. We must commit ourselves to a future where Americans and citizens of all nations will live and work in space.

And today, yes, the U.S. is the richest nation on Earth, with the most powerful economy in the world. And our goal is nothing less than to establish the United States as the preeminent spacefaring nation.

From the voyages of Columbus to the Oregon Trail to the journey to the Moon itself: history proves that we have never lost by pressing the limits of our frontiers. Indeed, earlier this month, one news magazine reported that Apollo paid down-to-earth dividends, declaring that man's conquest of the Moon "would have been a bargain at twice the price." And they called Apollo "the best return on investment since Leonardo da Vinci bought himself a sketch pad."

In 1961 it took a crisis—the space race—to speed things up. Today we don't have a crisis; we have an opportunity. To seize this opportunity, I'm not proposing a 10-year plan like Apollo; I'm proposing a long-range, continuing commitment. First, for the coming decade, for the 1990's: Space Station Freedom, our critical next step in all our space endeavors. And next, for the new century: Back to the Moon; back to the future. And this time, back to stay. And then a journey into tomorrow, a journey to another planet: a manned mission to Mars.

***

And to those who may shirk from the challenges ahead, or who doubt our chances of success, let me say this: To this day, the only footprints on the Moon are American footprints. The only flag on the Moon is an American flag. And the know-how that accomplished these feats is American know-how. What Americans dream, Americans can do. And 10 years from now, on the 30th anniversary of this extraordinary and astonishing flight, the way to honor the Apollo astronauts is not by calling them back to Washington for another round of tributes. It is to have Space Station Freedom up there, operational, and underway, a new bridge between the worlds and an investment in the growth, prosperity, and technological superiority of our nation . . .

Like them, and like Columbus, we dream of distant shores we've not yet seen. Why the Moon? Why Mars? Because it is humanity's destiny to strive, to seek, to find. And because it is America's destiny to lead.

***

When he announced his approval of the space station in January 1984, President Reagan had directed NASA to have the station in orbit "within a decade." By the time Bill Clinton was elected in November 1992, almost no space station hardware had been built. Freedom would certainly not be ready for service within two years. In fact, as Clinton began his first term as president in January 1993, his OMB director recommended canceling the struggling program, for which congressional support continued to wane. Rather than accept that recommendation, Clinton decided to order a major redesign of the space station. As that effort was getting under way, an unexpected message arrived at NASA. It was from the two most senior people in the Russian space program, Yuri Koptev, head of the new Russian Space Agency, and Yuri Semenyov, director of the major Russian space equipment manufacturer Energia.

After the collapse of the Soviet Union in 1991, Russia took control of most Soviet space capabilities, but with the post-Soviet economy in poor condition, there were few funds available for new space efforts. Koptev and Semenyov proposed what was in its essence a merger of the U.S. and Russian space station programs, Freedom *and Mir 2, the follow-on to the* Mir *station that Russia had launched in 1986.*

Letter from Yuri Koptev (General Director of the Russian Space Agency), and Yuri Semenyov (General Director and General Designer of NPO "Energia"), to NASA Administrator Daniel Goldin, March 16, 1993

Dear Mr. Goldin:

With great satisfaction we remember our meetings with you which output to broad cooperation between our countries in the area of space and in particular in the field of manned space flight. That is the area where Russia and USA have non-disputable priority.

Our country collected a broad record of developing and exploitation of manned multipurpose orbital complexes ("Salyut," "Mir"). In the meantime Russia developed conceptual design of a new generation orbital complex "Mir-2," which we plan to begin deployment in 1997. We are aware about all efforts of USA and its international partners to create space station "Freedom." Realization of such labor consuming projects demands beside all sophisticated scientific, technical and technological matters very significant financial resources to be spent. In an environment of increasing complexity of orbital manned structures and their objectives that becomes a serious problem pursuant to decrease the cost of R&D. In this direction there can be indisputable advantages which can be achieved by unification of efforts of Russia and USA in a course of realization of joint advanced orbital station with application of existent scientific, technical and constructional records.

We consider a possibility to suggest for your attention a program of international space station, which could imply the key project elements of space station "Mir" and "Freedom" and could provide for programmatic and economical profit for all participant countries.

Proposed concept is done below in annex and is based on the following principles:

- it includes a main core module of the orbital station "Mir-2" as basic element for upgrading;
- within following steps to the core module are to be added the US manned laboratory, the ESA laboratory "Columbus" and Japanese experimental module that could provide for building of really international orbital test bed;
- station operates on orbit with inclination over 50°, providing essential capabilities for Earth surface observation;
- this effort could be supported by various national launchers, including "Space Shuttle," "Soyuz," "Ariane-4,5."
- such orbital complex could provide initial capabilities for permanent manned presence of crew of three astronauts in 1997 with capability to increase it up to 9 in 2000 year.

Preliminary assessment showed that unification of technical capabilities and resources within the frame of this proposed international program could bring benefit of saving of some billions of dollars in spending to compare with now planned cost of realization for separate national programs for orbital stations development.

We consider that a proposed basis of this concept has vast potential for national and international interests. We are ready to represent and discuss with you this the most important matter in the nearest time frame which is convenient for you.

NASA administrator Dan Goldin, a holdover from the Bush administration, took the Russian proposal to the White House. President Clinton had deactivated the National Space Council at the start of his presidency, but his vice president, Al Gore, still took a lead role on space matters, and it was he and his staff who decided to give serious consideration to the Russian initiative, which was attractive to the White House in geopolitical terms. Among other advantages, it would be a way of signaling U.S. support for the more democratic government of Russian president Boris Yeltsin and of providing continued employment for Russian aerospace engineers who might otherwise seek jobs with countries such as Iran, North Korea, and other states hostile to the United States.

At an April 1993 get-acquainted meeting with Yeltsin, President Clinton indicated that the United States was inclined to accept the Russian proposal. After some difficult negotiations, on September 1, 1993, Gore and Russian prime minister Viktor Chernomyrdin signed an agreement to explore ways of jointly developing a space station.

The work of adding Russian hardware to a station already being redesigned began quickly. The United States also urged its partners in Freedom—several European countries working through the European Space Agency, Japan, and Canada—to extend along with the United States an invitation to Russia to join the space station partnership. That invitation was offered in December 1993. After four years of difficult negotiations, Russia and the original space station partners agreed in 1997 to work together on what by then was called the International Space Station (ISS). The first element of the ISS, a Russian module, was launched in 1998.

Other than this significant change in the space station program, there was little innovation in human spaceflight during the Clinton presidency. The NASA budget actually declined in purchasing power. The final statement of National Space Policy during the Reagan administration, issued in February 1988, had included as a long-term goal "expanding human presence and activity beyond Earth orbit into the solar system." The Clinton version of the policy, issued in 1996, removed that statement, saying only that "the International Space Station will support future decisions on the feasibility and desirability of conducting further human exploration activities." By the time Bill Clinton left the White House, the United States had no goal for human spaceflight beyond completing assembly of the International Space Station.

As George W. Bush became president in January 2001, that situation continued. NASA hoped to use the space shuttle to complete assembly of the U.S. elements of the International Space Station by 2004, then fly the shuttle on space station supply missions until 2020 or even beyond. There were no approved plans for humans to journey beyond the space station to more distant destinations.

Then, on February 1, 2003, space shuttle orbiter Columbia *disintegrated on its reentry from a sixteen-day science mission; like* Challenger, Columbia *was carrying a seven-person crew, and all died as the orbiter came apart. A Columbia Accident Investigation Board (CAIB) was immediately established to identify the causes of the accident and steps needed to return the shuttle to safe flight. (The author was a member of that thirteen-person board.)*

Unlike the situation following the Challenger *accident, when the Rogers Commission report limited itself to the physical causes of the accident and the steps needed to return the shuttle to flight, the CAIB delved into the underlying factors within NASA that had created the context leading to the accident. In its report, the board listed not only the physical cause of the accident but also its organizational roots.*

Columbia Accident Investigation Board, "Report," August 23, 2003, Executive Summary

The Columbia Accident Investigation Board's independent investigation into the February 1, 2003, loss of the Space Shuttle *Columbia* and its seven-member crew lasted nearly seven months. A staff of more than 120, along with some 400 NASA engineers, supported the Board's 13 members. Investigators examined more than 30,000 documents, conducted more than 200 formal interviews, heard testimony from dozens of expert witnesses, and reviewed more than 3,000 inputs from the general public. In addition, more than 25,000 searchers combed vast stretches of the Western United States to retrieve the spacecraft's debris. In the process, *Columbia*'s tragedy was compounded when two debris searchers with the U.S. Forest Service perished in a helicopter accident.

The Board recognized early on that the accident was probably not an anomalous, random event, but rather likely rooted to some degree in NASA's history and the human space flight program's culture. Accordingly, the Board broadened its mandate at the outset to include an investigation of a wide range of historical and organizational issues, including political and budgetary considerations, compromises, and changing priorities over the life of the Space Shuttle Program. The Board's conviction regarding the importance of these factors strengthened as the investigation progressed, with the result that this report, in its findings, conclusions, and recommendations, places as much weight on these causal factors as on the more easily understood and corrected physical cause of the accident.

The physical cause of the loss of *Columbia* and its crew was a breach in the Thermal Protection System on the leading edge of the left wing, caused by a piece of insulating foam which separated from the left bipod ramp section of the External Tank at 81.7 seconds after launch, and struck the wing in the vicinity of the lower half of Reinforced Carbon-Carbon panel number 8. During re-entry this breach in the Thermal Protection System allowed superheated air to penetrate through the leading edge insulation and progressively melt the aluminum structure of the left wing, resulting in a weakening of the structure until increasing aerodynamic forces caused loss of control, failure of the wing, and breakup of the Orbiter. This breakup occurred in a flight regime in which, given the current design of the Orbiter, there was no possibility for the crew to survive.

The organizational causes of this accident are rooted in the Space Shuttle Program's history and culture, including the original compromises that were required to gain approval for the Shuttle, subsequent years of resource constraints, fluctuating priorities, schedule pressures, mischaracterization of the Shuttle as operational rather than developmental, and lack of an agreed national vision for human space flight. Cultural traits and organizational practices detrimental to safety were allowed to develop, including: reliance on past success as a substitute for sound engineering practices (such as testing to understand why systems were not performing in accordance with requirements); organizational barriers that prevented effective communication of critical safety information and stifled professional differences of opinion; lack of integrated management across program elements; and the evolution of an informal chain of command and decision-making processes that operated outside the organization's rules.

***

The Bush White House paid close attention to chapter 9 of the CAIB report, "A Look Ahead," which pointed out "the lack, over the past three decades, of any national mandate providing NASA a compelling mission requiring human presence in space" and suggested that there had been "a failure of national leadership" in this respect. The CAIB also noted, "All members of the Board agree that America's future space efforts must include human presence in Earth orbit, and eventually beyond." Between September and December 2003, working with a few people from NASA, the White House put together a new plan for the future of U.S. human spaceflight. President Bush announced this "Vision for Space Exploration" in a speech at NASA Headquarters in Washington on January 14, 2004.

President George W. Bush, "Remarks at the National Aeronautics and Space Administration," January 14, 2004

***

Inspired by all that has come before and guided by clear objectives, today we set a new course for America's space program. We will give NASA a new focus and vision for future exploration. We will build new ships to carry man forward into the universe, to gain a new foothold on the moon, and to prepare for new journeys to worlds beyond our own.

***

Two centuries ago, Meriwether Lewis and William Clark left St. Louis to explore the new lands acquired in the Louisiana Purchase. They made that journey in the spirit of discovery, to learn the potential of vast new territory and to chart a way for others to follow. America has ventured forth into space for the same reasons. We have undertaken space travel because the desire to explore and understand is part of our character.

. . . Much remains for us to explore and to learn. In the past 30 years, no human being has set foot on another world or ventured farther upward into space than 386 miles, roughly the distance from Washington, DC, to Boston, Massachusetts. America has not developed a new vehicle to advance human exploration in space in nearly a quarter-century. It is time for America to take the next steps.

Today I announce a new plan to explore space and extend a human presence across our solar system. We will begin the effort quickly, using existing programs and personnel. We'll make steady progress, one mission, one voyage, one landing at a time.

Our first goal is to complete the International Space Station by 2010. We will finish what we have started. We will meet our obligations to our 15 international partners on this project . . .

To meet this goal, we will return the space shuttle to flight as soon as possible, consistent with safety concerns and the recommendations of the Columbia Accident Investigation Board. The shuttle's chief purpose over the next several years will be to help finish assembly of the International Space Station. In 2010, the space shuttle, after nearly 30 years of duty, will be retired from service.

Our second goal is to develop and test a new spacecraft, the crew exploration vehicle, by 2008 and to conduct the first manned mission no later than 2014. The crew exploration vehicle will be capable of ferrying astronauts and scientists to the space station after the shuttle is retired. But the main purpose of this spacecraft will be to carry astronauts beyond our orbit to other worlds. This will be the first spacecraft of its kind since the Apollo Command Module.

Our third goal is to return to the moon by 2020, as the launching point for missions beyond. Beginning no later than 2008, we will send a series of robotic missions to the lunar surface to research and prepare for future human exploration. Using the crew exploration vehicle, we will undertake extended human missions to the moon as early as 2015, with the goal of living and working there for increasingly extended periods of time. Eugene Cernan, who is with us today, the last man to set foot on the lunar surface, said this as he left, "We leave as we came, and God willing as we shall return, with peace and hope for all mankind." America will make those words come true.

Returning to the moon is an important step for our space program. Establishing an extended human presence on the moon could vastly reduce the costs of further space exploration, making possible ever more ambitious missions. Lifting heavy spacecraft and fuel out of the Earth's gravity is expensive. Spacecraft assembled and provisioned on the moon could escape its far lower gravity using far less energy and thus far less cost. Also, the moon is home to abundant resources. Its soil contains raw materials that might be harvested and processed into rocket fuel or breathable air. We can use our time on the moon to develop and test new approaches and technologies and systems that will allow us to function in other, more challenging environments. The moon is a logical step toward further progress and achievement.

With the experience and knowledge gained on the moon, we will then be ready to take the next steps of space exploration, human missions to Mars and to worlds beyond . . . The human thirst for knowledge ultimately cannot be satisfied by even the most vivid pictures or the most detailed measurements. We need to see and examine and touch for ourselves. And only human beings are capable of adapting to the inevitable uncertainties posed by space travel . . . We do not know where this journey will end, yet we know this: Human beings are headed into the cosmos.

***

It took more than a year for NASA to settle on a program to implement the Vision for Space Exploration. Michael Griffin became NASA administrator in April 2005 and by the end of that year had approved what came to be called the Constellation Program. Its initial elements included a crew exploration vehicle named Orion and two new launch vehicles, Ares-1 to launch

Orion and a Saturn V-class Ares-5 for lunar missions. Griffin called Orion "Apollo on steroids." The spacecraft would be much larger than the Apollo spacecraft, but like the Apollo command and service module (and unlike the space shuttle), it would be a capsule that at least initially would land in the ocean, where it and the crew would be recovered. A fourth part of the Constellation plan was a lunar landing vehicle called Altair.

NASA quickly began work on Orion and Ares-1, but the Bush administration did not provide the additional funding it had promised and there were various technical problems with the two systems. By the time the Bush administration was preparing to leave office after the 2008 election, there were a number of questions being raised about the viability of the Constellation Program.

After he was elected president in November 2008, Barack Obama appointed a NASA transition team which alerted him to the ongoing criticisms of the Constellation Program. The team suggested that his administration make its own assessment of Constellation before embracing it as the Obama effort in human spaceflight. Obama accepted that advice, and the White House in May 2009 created a "Review of U.S. Human Space Flight Plans" committee to make that assessment. In its October 2009 report, the committee concluded that "the U.S. human spaceflight program appears to be on an unsustainable trajectory." The committee's report was critical of the Constellation Program and recommended a number of changes to put U.S. human spaceflight on a more sustainable path.

The White House agreed with the committee's findings, and in February 2010 President Obama announced he was canceling Constellation, including the Orion spacecraft and the Ares family of launch vehicles. As a substitute for Orion as a means of transporting crew to and from the International Space Station, he proposed that NASA enter into a commercial partnership with two or more private firms that would develop crew transportation systems and provide transportation services to NASA and other customers. Obama also proposed greatly increased government investment in "game changing technologies," so that any new launch vehicles or other systems developed would be based on twenty-first-century technology such as advanced *rocket propulsion, artificial intelligence, and state-of-the-art electronics.*

There was immediate criticism of almost every aspect of the Obama strategy, which represented a significant change from "business as usual." A particular objection was the lack of any stated goal or destination for human spaceflight. To respond to this criticism and to defend his strategy, Barack Obama journeyed to the Kennedy Space Center in Florida on April 15, 2010. There he gave a speech in which he outlined his vision for a space effort that would carry humans to Mars orbit, and then to the Martian surface.

Barack Obama, "Remarks at the Kennedy Space Center," April 15, 2010

***

For me, the space program has always captured an essential part of what it means to be an American: reaching for new heights, stretching beyond what previously did not seem possible. And so as President, I believe that space exploration is not a luxury, it's not an afterthought in America's quest for a brighter future; it is an essential part of that quest.

So today I'd like to talk about the next chapter in this story. Now, the challenges facing our space program are different and our imperatives for this program are different than in decades past. We're no longer racing against an adversary. We're no longer competing to achieve a singular goal like reaching the Moon. In fact, what was once a global competition has long since become a global collaboration. But while the measure of our achievements has changed a great deal over the past 50 years, what we do—or fail to do—in seeking new frontiers is no less consequential for our future in space and here on Earth.

So let me start by being extremely clear: I am 100 percent committed to the mission of NASA and its future. Because broadening our capabilities in space will continue to serve our society in ways that we can scarcely imagine. Because exploration will once more inspire wonder in a new generation, sparking passions and launching careers. And because, ultimately, if we fail to press forward in the pursuit of discovery, we are ceding our future and we are ceding that essential element of the American character.

Now, I know there have been a number of questions raised about my administration's plan for space exploration, especially in this part of Florida, where so many rely on NASA as a source of income as well as a source of pride and community. And these questions come at a time of transition, as the space shuttle nears its scheduled retirement after almost 30 years of service. And understandably, this adds to the worries of folks concerned not only about their own futures, but about the future of the space program to which they've devoted their lives.

But I also know that underlying these concerns is a deeper worry, one that precedes not only this plan, but this administration. It stems from the sense that people in Washington, driven sometimes less by vision than by politics, have for years neglected NASA's mission and undermined the work of the professionals who fulfill it. We've seen that in the NASA budget, which has risen and fallen with the political winds.

But we can also see it in other ways: in the reluctance of those who hold office to set clear, achievable objectives, to provide the resources to meet those objectives, and to justify not just these plans, but the larger purpose of space exploration in the 21st century.

***

We will build on the good work already done on the Orion crew capsule. I've directed [NASA administrator] Charlie Bolden to immediately begin developing a rescue vehicle using this technology, so we are not forced to rely on foreign providers if it becomes necessary to quickly bring our people home from the International Space Station. And this Orion effort will be part of the technological foundation for advanced spacecraft to be used in future deep space missions. In fact, Orion will be readied for flight right here in this room.

Next, we will invest more than $3 billion to conduct research on an advanced heavy lift rocket, a vehicle to efficiently send into orbit the crew capsules, propulsion systems, and large quantities of supplies needed to reach deep space. Now, in developing this new vehicle, we will not only look at revising or modifying older models, we want to look at new designs, new materials, new technologies that will transform not just where we can go, but what we can do when we get there. And we will finalize a rocket design no later than 2015 and then begin to build it . . . I want everybody to understand, that's at least 2 years earlier than previously planned, and that's conservative, given that the previous program was behind schedule and over budget.

At the same time, after decades of neglect, we will increase investment right away in other groundbreaking technologies that will allow astronauts to reach space sooner and more often, to travel farther and faster for less cost, and to live and work in space for longer periods of time more safely . . .

So the point is, what we're looking for is not just to continue on the same path, we want to leap into the future. We want major breakthroughs, a transformative agenda for NASA.

***

Early in the next decade, a set of crewed flights will test and prove the systems required for exploration beyond low Earth orbit. And by 2025, we expect new spacecraft designed for long journeys to allow us to begin the first-ever crewed missions beyond the Moon into deep space. So we'll start by sending astronauts to an asteroid for the first time in history. By the mid-2030s, I believe we can send humans to orbit Mars and return them safely to Earth. And a landing on Mars will follow. And I expect to be around to see it.

***

Now, I understand that some believe that we should attempt a return to the surface of the Moon first, as previously planned. But I just have to say pretty bluntly here, we've been there before . . . There's a lot more of space to explore and a lot more to learn when we do. So I believe it's more important to ramp up our capabilities to reach and operate at a series of increasingly demanding targets, while advancing our technological capabilities with each step forward. And that's what this strategy does. And that's how we will ensure that our leadership in space is even stronger in this new century than it was in the last.

***

There was considerable opposition to the new NASA exploration strategy proposed by the Obama administration, in particular because it would result in the cancellation of contracts associated with the Constellation Program, with the associated loss of jobs. Members of Congress, particularly in the Senate, reacting to the threats to their constituents' jobs and possible lasting damage to the U.S. space industrial base, crafted an alternative approach to future human spaceflight that would require NASA to begin work immediately on a "multi-purpose crew vehicle" and a heavy lift launch vehicle, the Space Launch System. The technical characteristics of the Space Launch System were specified, and a schedule for spacecraft and launch vehicle development set out. It was unusual for members of Congress and their staffs to get so deep into the details of space program management. These developments were to be carried out as much as possible by modifying existing shuttle and Constellation contracts, thereby preserving existing jobs. The Senate approach was written into the 2010 NASA authorization bill.

Rather than fight for his new approach to space development, President Obama signed the NASA 2010 authorization bill into law in October 2010. The Space Launch System and the Orion Multi-Purpose Crew Launch Vehicle became NASA's major human spaceflight programs for the 2010s, as the first steps in what NASA characterized as a "Journey to Mars."

The 2010 NASA authorization bill also called for an independent assessment of the future of the U.S. human spaceflight effort by the National Academies of Sciences and Engineering, with particular attention to identifying a sustainable rationale for that effort. The operating arm of the academies, the National Research Council, established a Committee on Human Spaceflight to carry out this study. The committee's 2014 report, "Pathways to Exploration," provided a rather sober assessment of future prospects.

Committee on Human Spaceflight, National Research Council, "Pathways to Exploration: Rationales and Approaches for a U.S. Program of Human Space Exploration," 2014

1.7 SUMMARY: A SUSTAINABLE U.S. HUMAN SPACE EXPLORATION PROGRAM

Human space exploration requires a long-term commitment by the nation or entity that undertakes it. Therefore, the committee has concluded the following:

National leadership and a sustained consensus on the vision and goals are essential to the success of a human space exploration program that extends beyond LEO [low Earth orbit]. Frequent changes in the goals for U.S. human space exploration waste resources and impede progress. The instability of goals for the U.S. program of human spaceflight beyond LEO threatens our nation's appeal and suitability as an international partner.

The United States has had a sustained program of human spaceflight for more than a half-century, paradoxically in the face of, at best, lukewarm public support. There has not been a committed, passionate minority large and influential enough to maintain momentum for the kind of dramatic progress that was predicted by many space experts at the time of Apollo. That is a problem that adds to the numerous difficulties—frequent redirection, mismatch of mission and resources, and political micromanagement—that have afflicted the U.S. human spaceflight program since Apollo ended. The committee has concluded as follows:

Simply setting a policy goal is not sufficient for a sustainable human spaceflight program, because policy goals do not change programmatic, technical, and budgetary realities. Those who are formulating policy goals need to keep the following factors in mind:

- No defensible calculation of tangible, quantifiable benefits—spinoff technologies, attraction of talent to scientific careers, scientific knowledge, and so on—is likely ever to demonstrate a positive return on the massive investment required by human spaceflight.

- The arguments that triggered the Apollo investment—national defense and prestige—seem to have especially limited public salience in today's post–Cold War America.
- Although the public is mostly positive about NASA and its spaceflight programs, increased spending on spaceflight has low priority for most Americans. However, most Americans do not follow the issue closely, and those who pay more attention are more supportive of space exploration.

It serves no purpose for advocates of human exploration to dismiss those realities in an era in which both the citizenry and national leaders are focused intensely on the unsustainability of the national debt, the dramatic growth of entitlement spending, and the consequent downward pressure on discretionary spending, including the NASA budget. With most projections forecasting growing national debt in the decades ahead, there is at least as great a chance that human spaceflight budgets will be below the recent flat trend line as that they will be markedly above it.

Nevertheless, the committee has concluded as follows:

If the United States decides that the intangible benefits of human spaceflight justify major new and enduring public investments in human spaceflight, it will need to craft a long-term strategy that will be robust in the face of technical and fiscal challenges.

***

None of those steps can replace the element of sustained commitment on the part of those who govern the nation, without which neither Apollo nor its successor programs would have occurred. Hard as the above choices may appear, they probably are less difficult or less alien for conventional political decision-makers than the recognition that human spaceflight—among the longest-term of long-term endeavors—cannot be successful if held hostage to traditional short-term decision-making and budgetary processes. Asking future presidents to preserve rather than tinker with previously chosen pathways or asking Congresses present and future to fund human spaceflight aggressively with budgets that increase by more than the rate of inflation every year for decades may seem fanciful. But it is no less so than imagining a magic rationale that ignites and then sustains a public demand that has never existed in the first place. Americans have continued to fly into space not so much because the public strongly wants it to be so but because the counterfactual—space exploration dominated by the vehicles and astronauts of other nations—seems unthinkable after 50 years of U.S. leadership in space. In reviving a U.S. human exploration program capable of answering the enduring questions about humanity's destiny beyond our tiny blue planet, we will need to grapple with the attitudinal and fiscal realities of the nation today while staying true to a small but crucial set of fundamental principles for the conduct of exploration of the endless frontier.

The election of Donald Trump as president in November 2016 once again changed the goals of the U.S. human spaceflight effort. Trump reactivated the National Space Council, with his vice president, Mike Pence, as its chair. At the initial meeting of the Space Council on October 5, 2017, Pence declared, "We will return American astronauts to the moon, not only to leave behind footprints and flags, but to build the foundation we need to send Americans to Mars and beyond." Two months later, President Trump signed an executive order modifying the Obama policy of skipping the Moon on the way to Mars and directing NASA to "lead an innovative and sustainable program of exploration."

Since 2004 the United States, under three presidents, has moved in fits and starts with respect to pursuing a space program intended to once again send Americans away from their home planet. Whether that effort will succeed, and whether the United States will take a leading position in twenty-first-century space exploration, remains to be seen.

President Donald Trump, "Space Policy Directive 1," December 11, 2017

SUBJECT: REINVIGORATING AMERICA'S HUMAN SPACE EXPLORATION PROGRAM

Section 1. Amendment to Presidential Policy Directive-4.

Presidential Policy Directive-4 of June 28, 2010 (National Space Policy), is amended as follows:

The paragraph beginning "Set far-reaching exploration milestones" is deleted and replaced with the following:

"Lead an innovative and sustainable program of exploration with commercial and international partners to enable human expansion across the solar system and to bring back to Earth new knowledge and opportunities. Beginning with missions beyond low-Earth orbit, the United States will lead the return of humans to the Moon for long-term exploration and utilization, followed by human missions to Mars and other destinations."

***

THE DREAM PERSISTS

In the first decades of the twenty-first century, the future of human spaceflight sponsored by the U.S. government has become uncertain. While rhetorically recent presidents and Congresses have supported resuming human spaceflight beyond Earth orbit, their actions, particularly with respect to providing the funding required to meet the goals they have set, have not always kept pace with their words. Missing so far is the sustained commitment that would allow NASA to take the next steps in a long-term program aimed at sending humans beyond the immediate vicinity of their home planet.

In recent years, however, an alternative to government-sponsored spaceflight has emerged in the United States. This alternative is the decision of several extremely wealthy individuals to create companies to make real the promise of privately funded human spaceflight into Earth orbit and beyond. The earliest of these ventures was Blue Origin, founded by Jeff Bezos of Amazon in 2000. Blue Origin was followed by Space Exploration Technologies, generally known as SpaceX, founded by internet billionaire Elon Musk in 2002.

Blue Origin's motto is "*Gradatim Ferociter*," Latin for "Step by Step, Ferociously." Its website declares, "Earth, in all its beauty, is just our starting place." In recent years, Bezos has identified his goal for Blue Origin as making it possible to have "millions of people living and working in space, and exploring the entire solar system." Because of the abundance of natural resources in asteroids and comets, and the ease of generating power in orbit, he believes most heavy industries will migrate to space, leaving Earth protected from their environmental impacts. Though Blue Origin has successfully sent unmanned vehicles into space, Bezos has not articulated his long-term strategy for achieving these goals, and the company is some years behind SpaceX in developing the capabilities required to send humans into orbit and beyond.

By contrast, SpaceX has since its first years actively pursued opportunities to become involved in current space activities in both the public and private sectors. It has developed a workhorse launch vehicle, the Falcon 9, with a heavy lift version, Falcon Heavy, first launched on February 6, 2018, and has marketed its launch capabilities with significant success. It has been able to recover, refurbish, and reuse the first stages of a number of the Falcon 9 boosters, enabling it to set launch prices significantly lower than those of its competitors. SpaceX has also developed a spacecraft called Dragon to carry supplies and eventually humans to the International Space Station. The SpaceX website declares that "the company was founded in 2002 to revolutionize space technology, with the ultimate goal of enabling people to live on other planets."

It may well be that people like Jeff Bezos and Elon Musk—or others to come—are the twenty-first-century successors of Wernher von Braun, combining an expansive vision of the future of human spaceflight with an ability to turn that vision into reality. While von Braun advanced his vision by persuading governments (including Nazi Germany) to embrace some of its elements, Bezos and Musk seem intent on using private resources to the greatest degree possible in fostering future human spaceflight.

It is thus fitting that this volume, which began with a visionary 1954 essay by Wernher von Braun about a human mission to Mars, concludes with the vision laid out in September 2016 by Elon Musk of a plan to create a million-person city on Mars. Much of what von Braun set out more than sixty years ago became reality, though not yet a human mission to Mars. Whether it will be SpaceX, Blue Origin, various governments, or some combination of those actors that takes that next "great leap" into space is yet to be determined.

At the September 2016 International Astronautical Congress in Guadalajara, Mexico, SpaceX chief executive officer Elon Musk set out a bold vision for future human expansion beyond Earth, in particular by establishing a million-person city on the

*surface of Mars. At the same congress a year later, in Adelaide,
Australia, he modified his 2016 vision to be slightly more modest. This essay is an edited version of the transcript of Musk's
2016 talk; the transcript originally appeared in the June 2017
issue of the journal* New Space *and is used with permission from
both the journal and SpaceX.*

Elon Musk (Chief Executive Officer, Space X), "Making Humans a Multi-Planetary Species," September 2016

By talking about the SpaceX Mars architecture, I want to make
Mars seem possible—make it seem as though it is something that
we can do in our lifetime. There really is a way that anyone could
go if they wanted to.

WHY GO ANYWHERE?

I think there are really two fundamental paths to the future. History is going to bifurcate along two directions. One path is we
stay on Earth forever, and then there will be some eventual ex-
tinction event. I do not have an immediate doomsday prophecy,
but eventually, history suggests, there will be some doomsday
event. The alternative is to become a space-faring civilization and
a multi-planetary species, which I hope you would agree is the
right way to go.

So how do we figure out how to take you to Mars and create a
self-sustaining city—a city that is not merely an outpost but
which can become a planet in its own right, allowing us to become a truly multi-planetary species?

WHY MARS?

Sometimes people wonder, "Well, what about other places in the
solar system? Why Mars?" Our options for becoming a multi-
planetary species within our solar system are limited. We have,
in terms of nearby options, Venus, but Venus is a super-high-
pressure hot acid bath, so that would be a tricky one. Venus is
not at all like the goddess. So, it would be really difficult to make
things work on Venus. Then, there is Mercury, but that is way
too close to the sun. We could potentially go onto one of the

moons of Jupiter or Saturn, but those are quite far out, much further from the sun, and much harder to get to.

This really only leaves us with one option if we want to become a multi-planetary civilization, and that is Mars. We could conceivably go to our moon, and I actually have nothing against going to the moon, but I think it is challenging to become multiplanetary on the moon because it is much smaller than a planet. It does not have any atmosphere. It is not as resource-rich as Mars. It has got a 28-day day, whereas the Mars day is 24.5 hours. In general, Mars is far better-suited ultimately to scale up to be a self-sustaining civilization.

The two planets are remarkably close in many ways (See Table 1). In fact, we now believe that early Mars was a lot like Earth. In effect, if we could warm Mars up, we would once again have a thick atmosphere and liquid oceans. Mars is about half as far again from the sun as Earth is, so it still has decent sunlight. It is a little cold, but we can warm it up. It has a very helpful atmosphere, which, being primarily CO_2 with some nitrogen and argon and a few other trace elements, means that we can grow plants on Mars just by compressing the atmosphere.

Table 1. Characteristics of the Earth and Mars

	Earth	Mars
Diameter	12,756 km/7,926 mi	6,792 km/4,220 mi
Average distance from sun	150,000,000 km/ 93,000,000 mi	229,000,000 km/ 142,000,000 mi
Temperature range	−88°C to 58°C/−126°F to 138°F	−140°C to 30°C/− 285°F to 88°F
Atmospheric composition	78% N_2, 21% O_2, 1% other	96% CO_2, < 2% Ar, <2% N_2, < 1% other
Force of gravity (weight)	100 lbs on Earth	38 lbs on Mars (62.5% less gravity)
Day length	24 hrs	24 hrs 40 min
Land mass	148.9 million km²	144.8 million km² (97% of Earth)
People	7 billion	0

It would be quite fun to be on Mars because you would have gravity that is about 37% of that of Earth, so you would be able to lift heavy things and bound around. Furthermore, the day is remarkably close to that of Earth. We just need to change the populations, because currently we have seven billion people on Earth and none on Mars.

FROM EARLY EXPLORATION TO A SELF-SUSTAINING CITY ON MARS

There has been a lot of great work by NASA and other organizations in the early exploration of Mars and understanding what Mars is like. Where could we land? What is the composition of the atmosphere? Where is there water or ice? We need to go from these early exploration missions to actually building a city.

The issue that we have today is that there is no overlap between people who want to go to Mars and those who can afford to go. In fact, right now, you cannot go to Mars even if you had infinite money.

Using traditional methods, taking an Apollo-style approach to get humans to Mars, an optimistic cost would be about $10 billion per person. The cost estimates for Apollo are somewhere between $100 and $200 billion in current-year dollars, and we sent 12 people to the surface of the moon, which was an incredible thing—probably one of the greatest achievements of humanity. However, that is a steep price to pay for a ticket. You cannot create a self-sustaining civilization if the ticket price is $10 billion per person.

What we need to do is to create some overlap between those who want to go and those who can afford to go. If we can get the cost of moving to Mars to be roughly equivalent to a median house price in the United States, which is around $200,000, then I think the probability of establishing a self-sustaining civilization is very high. I think it would almost certainly occur.

Not everyone would want to go. In fact, probably a relatively small number of people from Earth would want to go, but enough would want to go who could afford it for it to happen. People could also get sponsorship. It gets to the point where almost anyone, if they saved up and this was their goal, could buy a ticket and move to Mars—and given that Mars would have a labor shortage for a long time, jobs would not be in short supply.

IMPROVING COST PER TON TO MARS BY FIVE MILLION PERCENT

To get to this price point, we have to figure out how to improve the cost of trips to Mars by five million percent. This translates to an improvement of approximately four-and-a-half orders of magnitude. This is not easy. It sounds virtually impossible, but there are ways to do it.

There are four key elements that are needed in order to achieve the four-and-a-half orders of magnitude improvement. Most of the improvement would come from full reusability—somewhere between two and two-and-a-half orders of magnitude. The other two orders of magnitude would come from fueling in orbit, propellant production on Mars, and choosing the right propellant.

Full Reusability

To make Mars trips possible on a large enough scale to create a self-sustaining city, full reusability is essential. Full reusability is really the super-hard one. It is very difficult to achieve reusability even for an orbital system, and that challenge becomes substantially greater for a system that has to go to another planet.

You could use any form of transport as an example of the difference between reusability and expendability. A car, bicycle, horse, if they were single-use, almost no one would use them; it would be too expensive. However, with full reusability and frequent flights, you can take an aircraft that costs $90 million and buy a ticket on Southwest right now from Los Angeles to Vegas for $43, including taxes. If it were single use, it would cost $500,000 per flight. Right there, you can see an improvement of four orders of magnitude.

Now, applying reusability to trips to Mars is harder—reusability does not apply quite as much to Mars because the number of times that you can reuse the spaceship part of the system is less often than on Earth because the Earth to Mars launch window only occurs every 26 months. Therefore, you get to use the spaceship part only approximately every 2 years.

Fueling in Orbit

However, you could get to use the booster and the tanker parts more frequently. Therefore, it makes sense to send the spaceship into orbit with essentially its tanks dry. If it has really big tanks, and you can use the booster and tanker to fill them with fuel and oxidizer once in orbit, you can maximize the payload of the spaceship, so when it goes to Mars, you have a very large payload capability.

Fueling in orbit is one of the essential elements of this plan. Without fueling in orbit, you would have roughly a half order of magnitude impact on the cost. Each order of magnitude is a factor of 10. Therefore, not fueling in orbit would mean roughly a 500% increase in the cost per ticket.

Fueling in orbit also allows us to build a smaller vehicle and lower the development cost, although the vehicle is still quite big. However, it would be much harder to build something that is 5–10 times the size. Furthermore, it reduces the sensitivity of the performance characteristics of the booster rocket and tanker. So, if there is a shortfall in the performance of any of the elements, you can make up for it by having one or two extra fueling trips to the spaceship. This is very important for reducing the susceptibility of the system to a performance shortfall.

Propellant Production on Mars

Producing propellant on Mars is obviously also very important. Again, if we did not do this, it would have at least a half order of magnitude increase in the cost of a trip. It would be pretty absurd to try to build a city on Mars if your spaceships just stayed on Mars and did not go back to Earth. You would have a massive graveyard of ships; you would have to do something with them.

It would not really make sense to leave your spaceships on Mars; you would want to build a propellant plant on Mars and send the ships back. Mars happens to work out well for that because it has a CO_2 atmosphere, it has water ice in the soil, and with H_2O and CO_2, you can produce methane (CH_4) and oxygen (O_2).

Right Propellant

Picking the right propellant is also important. There are three main choices, and they each have their merits. First, there is kerosene, or rocket propellant-grade kerosene, essentially a highly refined form of jet fuel. It helps keep the vehicle size small, but because it is a very specialized form of jet fuel, it is quite expensive. Its reusability potential is lower. It would be very difficult to make this on Mars because there is no oil. Propellant transfer is pretty good but not great.

Hydrogen, although it has a high specific impulse, is very expensive, and it is incredibly difficult to keep from boiling off because liquid hydrogen is very close to absolute zero as a liquid. Therefore, the installation required is tremendous, and the energy cost on Mars of producing and storing hydrogen would be very high.

When we looked at the overall system optimization, it was clear that methane was the clear winner. Methane would require from 50% to 60% of the energy on Mars to refill propellant using the propellant depot, and the technical challenges are a lot easier. We therefore think methane is better almost across the board. We started off initially thinking that hydrogen would make sense, but ultimately came to the conclusion that the best way to optimize the cost-per-unit mass to Mars and back is to use an all-methane system—or technically, deep-cryo methalox.

Whatever system is designed, whether by SpaceX or someone else, these are the four features that need to be addressed in order for the system really to achieve a low cost per ton to the surface of Mars.

SYSTEM ARCHITECTURE

Figure 1 illustrates the overall system. The rocket booster and the spaceship take off and launch the spaceship into orbit. The rocket booster then comes back quite quickly, within about 20 minutes. So, it can actually launch the tanker version of the spacecraft, which is essentially the same as the spaceship but with the unpressurized and pressurized cargo areas filled up with propellant tanks. This also helps lower the development cost, which obviously will not be small.

Then, the propellant tanker goes up anywhere from three to five times to fill the tanks of the spaceship in orbit. Once the tanks are full, the cargo has been transferred, and we reach the Mars rendezvous timing, which is roughly every 26 months, that is when the ship would depart.

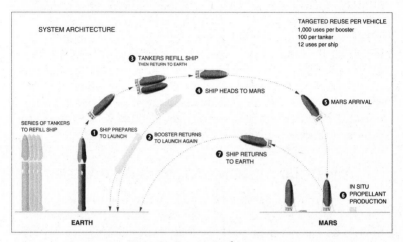

Figure 1. System Architecture.

Over time, there were would be many spaceships. You would ultimately have upwards of 1,000 or more spaceships waiting in orbit. Hence, the Mars Colonial Fleet would depart en masse.

It makes sense to fuel the spaceships in orbit because you have got 2 years to do so, and then you can make frequent use of the booster and the tanker to get really heavy reuse out of those. With the spaceship, you get less reuse because you have to consider how long it is going to last—maybe 30 years, which might be perhaps 12–15 flights of the spaceship at most.

Therefore, you really want to maximize the cargo of the spaceship and reuse the booster and the tanker as much as possible. Hence, the ship goes to Mars, gets replenished, and then returns to Earth. This ship will be relatively small compared with the Mars interplanetary ships of the future. However, it needs to fit 100 people or thereabouts in the pressurized section, carry the luggage and all of the unpressurized cargo to build propellant plants, and to build everything from iron foundries to pizza joints to you name it—we need to carry a lot of cargo.

The threshold for a self-sustaining city on Mars or a civilization would be a million people. If you can only go every 2 years and if you have 100 people per ship, that is 10,000 trips. Therefore, at least 100 people per trip is the right order of magnitude, and we may end up expanding the crew section and ultimately taking more like 200 or more people per flight in order to reduce the cost per person.

However, 10,000 flights is a lot of flights, so ultimately you would really want in the order of 1,000 ships. It would take a while to build up to 1,000 ships. How long it would take to reach that million-person threshold from the point at which the first ship goes to Mars would probably be somewhere between 20 and 50 total Mars rendezvous—so it would take 40–100 years to achieve a fully self-sustaining civilization on Mars.

VEHICLE DESIGN AND PERFORMANCE

The design of the spaceship is, in some ways, not all that complicated. The spaceship is made primarily of an advanced carbon fiber. The carbon-fiber part is tricky when dealing with deep cryogens and trying to achieve both liquid and gas impermeability and not have gaps occur due to cracking or pressurization that would make the carbon fiber leaky. Hence, it is a fairly significant technical challenge to make deeply cryogenic tanks out of carbon fiber. It is only recently that carbon-fiber technology has reached the point where we can do this without having to create a liner on the inside of the tanks, which would add mass and complexity.

This is particularly tricky for the pressurization of the hot gases. This is likely to be autogenously pressurized, which means that we gasify the fuel and the oxygen through heat exchanges in the engine and use that to pressurize the tanks. So, we gasify the methane and use that to pressurize the fuel tank, and we gasify the oxygen and use that to pressurize the oxygen tank.

This is a much simpler system than what we have with Falcon where we use helium for pressurization and nitrogen for gas thrusters. In this case, we would autogenously pressurize and then use gaseous methane and oxygen for the control thrusters. Hence, you really only need two ingredients for this, as opposed to four in the case of Falcon 9, or five if you consider the ignition liquid. In this case, we would use spark ignition.

For the expendable mode, the vehicle that we are proposing would lift about 550 tons into low orbit; it would lift about 300 tons in reusable mode. That compares to the Saturn V maximum capability of 135 tons. With most rockets that are currently flying, including ours, the weight-lifting capability is only a small percentage of the actual size of the rocket. However, with the interplanetary system which will initially be used for Mars, we believe we have improved the design performance massively. It is the first time a rocket lift capability will actually exceed the physical size of the rocket. The thrust level of the Mars launch vehicle is enormous. We are talking about a lift-off thrust of 13,000 tons, so it will be quite tectonic when it takes off. However, the booster does fit on Pad 39A at Kennedy Space Center, which NASA has been kind enough to allow us to use. NASA oversized the pad for Saturn V during Apollo. As a result, we can use a much larger vehicle on that same launch pad. In the future, we expect to add additional launch locations, probably adding one on the south coast of Texas.

The Saturn V and the Mars vehicle have very different purposes. The Saturn V was intended to be just powerful enough to send astronauts to the Moon in one launch. The Mars vehicle is really intended to carry huge numbers of people and ultimately millions of tons of cargo to Mars. Therefore, you really need something quite large in order to do that.

RAPTOR ENGINE

We started have development with what are probably the two most difficult key elements of the design of the interplanetary spaceship, the engine and rocket booster. The Raptor engine is going to be the highest chamber pressure engine of any kind ever built, and probably the highest thrust-to-weight.

Raptor is a full-flow staged combustion engine, which maximizes the theoretical momentum that you can get out of a given source fuel and oxidizer. We subcool the oxygen and methane to densify it. In most rockets, fuel and oxidizer are used close to their boiling points; in our case, we load the propellants close to their freezing point. That can result in a density improvement of around 10%–12%, which makes an enormous difference in the actual result of the rocket. It gets rid of any cavitation risk for the turbo pumps, and it makes it easier to feed a high-pressure turbo pump if you have very cold propellant.

One of the keys here, though, is the vacuum version of the Raptor having a 382-second ISP. This is critical to the whole Mars mission and we are confident we can get to that number or at least within a few seconds of that number, ultimately maybe even exceeding it slightly.

ROCKET BOOSTER

In many ways, the rocket booster is really a scaled-up version of the Falcon 9 booster. There are a lot of similarities, such as the grid fins and clustering a lot of engines at the base. The big differences are that the primary structure is an advanced form of carbon fiber as opposed to aluminum lithium, that we use autogenous pressurization, and that we get rid of the helium and the nitrogen.

Each rocket booster uses 42 Raptor engines. That is a lot of engines, but with Falcon Heavy, there are 27 engines on its base. Therefore, we will have considerable experience with a large number of engines. The large number also gives us redundancy, so that if some of the engines fail, you can still continue the mission and everything will be fine. The main job of the booster is to accelerate the spaceship to around 8,500 km/h. Orbital dynamics are all about velocity and not about altitude.

In the case of other planets, though, which have a gravity well that is not as deep, such as Mars, the moons of Jupiter, conceivably one day maybe even Venus—well, Venus will be a little trickier—but for most of the solar system, you only need the spaceship to land and take off. You do not need the rocket booster if you have a lower gravity well. No booster is needed on the moon or Mars or any of the moons of Jupiter, Saturn, or Pluto. The booster is just there for heavy gravity wells.

We have also been able to optimize the residual propellant needed for boost back and landing to get it down to about 7% of the lift-off propellant load. With some optimization, maybe we can get it down to about 6%. We are also now getting quite comfortable with the accuracy of the landing. If you have been watching the Falcon 9 landings, you will see that they are getting increasingly closer to the bull's eye. In particular, with the addition of maneuvering thrusters, we think we can actually put the booster right back on the launch stand. The fins at the base are thus essentially centering features to take out any minor position mismatch at the launch site.

We think we only need to gimbal or steer the center cluster of booster engines. There are seven engines in the center cluster. Those would be the ones that move for steering the rocket, and the other ones would be fixed in position. We can max out the number of engines because we do not have to leave any room for gimbaling or moving the other engines. This is all designed so that you could actually lose multiple engines, even at liftoff or anywhere in flight, and still continue the mission safely.

INTERPLANETARY SPACESHIP

On the top of the spaceship is the pressurized compartment. Beneath that is where we would have the unpressurized cargo, which would be really flat-packed—a very dense format. Below that is the liquid oxygen tank.

The liquid oxygen tank is probably the hardest piece of this whole vehicle to develop because it must handle propellant at the coldest level and the tanks themselves actually form the airframe. The airframe structure and the tank structure are combined, as is the case in all modern rockets. In aircraft, for example, the wing is really a fuel tank in the shape of a wing. The oxygen tank has to take the thrust loads of ascent and the loads of reentry, and it has to be impermeable to gaseous oxygen, which is tricky, and nonreactive to gaseous oxygen. These requirements make the tank the most difficult piece of the spaceship to construct, which is why we have already started on that element.

Below the oxygen tank is the fuel tank, and then the engines are mounted directly to the thrust cone on the base. There are six high-efficiency vacuum engines around the perimeter, and those do not gimbal. There are three of the sea-level versions of the engine, which do gimbal and provide the steering, although we can do some amount of steering when you are in space with differential thrust on the outside engines.

The net effect is a cargo to Mars of up to 450 tons, depending upon how many fills you do with the tanker. The goal is at least 100 passengers per ship, although ultimately we will probably see that number grow to 200 or more.

Depending upon which Earth-Mars rendezvous you are aiming for, the trip time at 6 km/s departure velocity can be as low as 80 days. Over time, we would improve that, and, eventually, I suspect that you would see Mars transit times of as little as 30 days in the more distant future. The duration of the journey is fairly manageable, considering the trips that people used to do in the old days, when sailing voyages would take 6 months or more.

For safe arrival, the heat-shield technology is extremely important. We have been refining the heat-shield technology using our Dragon spacecraft, and we are on version 3 of PICA, which is a phenolic-impregnated carbon ablator, and it is getting more robust with each new version, with less ablation, more resistance, and less need for refurbishment. The heat shield is basically a giant brake pad. It is a matter of how good you can make that brake pad against extreme re-entry conditions, while minimizing the cost of refurbishment, and eventually making it so that you could have many flights with no refurbishment at all.

I want to give you a sense of what it would feel like to actually be in the spaceship. In order to make the journey appealing and increase the number of people who actually want to go, it has got to be really fun and exciting—it cannot feel cramped or boring. Therefore, the crew compartment or the occupant compartment is set up so that you can do zero-gravity games—you can float around. There will be movies, lecture halls, cabins, and a restaurant. It will be really fun to go. You are going to have a great time!

PROPELLANT PLANT

The ingredients are there on Mars to create a propellant plant with relative ease because the atmosphere is primarily CO_2, and water-ice is almost everywhere. There is CO_2 plus H_2O to make methane, CH_4, and oxygen, O_2, using the Sabatier reaction. The trickiest thing really is the energy source, which can be done with a large field of solar panels.

COST PER TRIP

The key is making this trip affordable to almost anyone who wants to go. Based on this architecture, assuming optimization over time, we are looking at a cost per ticket of <$200,000, maybe as little as $100,000 over time, depending upon how much mass a person takes.

Right now, we are estimating about $140,000 per ton for the trips to Mars. If a person plus their luggage is less than that, taking into account food consumption and life support, the cost of moving to Mars could ultimately drop below $100,000.

Obviously, it is going to be a challenge to fund this whole endeavor. We expect to generate a pretty decent net cash flow from launching lots of satellites and servicing the space station for NASA, transferring cargo to and from the space station. There are also many people in the private sector who are interested in helping to fund a base on Mars, and perhaps there will be interest on the government sector side to do that too. Ultimately, this is going to be a huge public-private partnership.

We are now just trying to make as much progress as we can with the resources that we have available and to keep the ball moving forward. As we show that this is possible and that this dream is real—it is not just a dream, it is something that can be made real—the support will snowball over time.

I should also add that the main reason I am personally accumulating assets is in order to fund this. I really do not have any other motivation for personally accumulating assets except to be able to make the biggest contribution I can to making life multiplanetary.

TIMELINES

In 2002, SpaceX basically consisted of carpet and a mariachi band. That was it. I thought we had maybe a 10% chance of doing anything—of even getting a rocket to orbit, let alone getting beyond that and taking Mars seriously. However, I came to the conclusion that if there were no new entrants into the space arena with a strong ideological motivation, then it did not seem as if we were on a trajectory to ever be a space-based civilization and be out there among the stars.

In 1969, we were able to go to the moon, and after that the space shuttle could get to low Earth orbit. Then the space shuttle was retired. Since then, the trend line of how far into space humans could travel (at least on U.S. spacecraft) is down to zero. What many people do not appreciate is that technology does not automatically improve; it only improves if a lot of really strong engineering talent is applied to the problem. There are many examples in history where civilizations have reached a certain technology level, fallen well below that, and then recovered only millennia later.

SpaceX went from 2002, where we basically were clueless, and we built the smallest useful orbital rocket that we could think of with Falcon 1, which would deliver half a ton to orbit. Four years later, we developed the first vehicle. We developed the main engine, the upper-stage engine, the airframes, the fairing, and the launch system, and we had our first attempt at launch in 2006, which failed. The flight lasted only about 60 seconds, unfortunately.

However, in 2006, 4 years after starting, was also when we got our first NASA contract. I am incredibly grateful to NASA for supporting SpaceX, despite the fact that our rocket crashed. I am NASA's biggest fan. Thank you very much to the people who had the faith to do that.

Finally, the fourth launch of Falcon 1 worked in 2008. We were really down to our last pennies. In fact, I only thought I had enough money for three launches, and the first three failed! We were able to scrape together enough to just make it and do a fourth launch, and, thank goodness—that fourth launch succeeded in 2008. That time involved a lot of pain.

The end of 2008 is also when NASA awarded us the first major operational contract, which was for resupplying cargo to the space station and bringing cargo back. A couple of years later, we did the first launch of Falcon 9, version 1, and that had about a 10-ton-to-orbit capability, which was about 20 times the capability of Falcon 1. It was also assigned to carry our Dragon spacecraft.

It was in 2012 when we delivered and returned cargo from the space station. In 2013, we started doing vertical take-off and landing tests for the first time. Then, in 2014, we were able to have the first orbital booster do a soft landing in the ocean. The landing was soft, it fell over, and it exploded. However, for 7 seconds, the landing was good. We also improved the capability of the vehicle from 10 tons to about 13 tons to LEO. December 2015 was definitely one of the best moments of my life: the rocket booster came back and landed at Cape Canaveral. That really showed that we could bring an orbital-class booster back from a very high velocity, all the way to the launch site, and land it safely with almost no refurbishment required for reflight. If things go well, we are hoping to refly one of the landed boosters in a few months. (Editor's note: The first reflight took place in March 2017.)

In 2016, we also demonstrated landing on a ship, which is important for both the very high-velocity geosynchronous missions and for the reusability of Falcon 9, because about roughly a quarter of our missions service the space station. There are a few other lower-orbit missions, but probably 60% of our missions are commercial GEO missions. These high-velocity missions need to land on a ship out at sea. They do not have enough propellant onboard to boost back to the launch site.

FUTURE

Figure 2 shows the future—the next steps. We are intentionally fuzzy about this timeline. However, we are going to try to make as much progress as we can on a very constrained budget on the elements of the interplanetary transport booster and spaceship. Hopefully, we will be able to complete the first development spaceship in maybe about 4 years, and we will start doing suborbital flights with that.

That development vehicle actually has enough capability that you could possibly go to orbit if you limit the amount of cargo on the spaceship. You would have to really strip it down, but in tanker form, it could definitely get to orbit. It cannot get back, but it can get to orbit. Maybe there is some market for the really fast transport of things around the world, provided we can land somewhere where noise is not a super-big deal, because rockets are very noisy. We could transport cargo to anywhere on Earth in

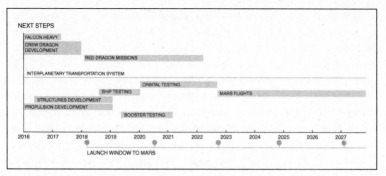

Figure 2. Next Steps in the Development of the Interplanetary Transportation System.

45 minutes at the most. Hence, most places on Earth would be 20–25 minutes away. If we had a floating platform off the coast of New York, 20–30 miles out, you could go from New York to Tokyo in 25 minutes and across the Atlantic in 10 minutes. Most of your time would be spent getting to the ship, and then it would be very quick after that. Therefore, there are some intriguing possibilities there, although we are not counting on that.

Then, there is the development of the booster. The booster part is relatively straightforward because it amounts to a scaling up of the Falcon 9 booster. So, we do not see that there will be many showstoppers there.

Next we will be trying to put it all together and make this actually work for Mars. If things go super-well, it might be in the 10-year timeframe, but I do not want to say that this is when it will occur. There is a huge amount of risk. It is going to cost a lot. There is a good chance we will not succeed, but we are going to do our best and try to make as much progress as possible.

RAPTOR FIRING

I was really excited to see all our Raptor engines firing. The Raptor is a really tricky engine. It is a lot trickier than a Merlin because it is a full-flow stage combustion, with much higher pressure. I am amazed that it did not blow up on the first firing, but fortunately it was good.

Although the Raptor has three times the thrust of a Merlin, it is actually only about the same size as a Merlin engine because it has three times the operating pressure. Part of the reason for making the engine small is so that we can use many of the production techniques that we honed with Merlin.

We are currently producing Merlin engines at almost 300 per year. Therefore, we understand how to make rocket engines in volume. Hence, even though the Mars vehicle uses 42 on the base and nine on the upper stage—so we have 51 engines to make—that is well within our production capabilities for Merlin. This is a similarly sized engine to Merlin, except for the expansion ratio. We therefore feel really comfortable about being able to make this engine in volume at a price that does not break our budget.

CARBON-FIBER TANK

We also wanted to make progress on the primary structure and particularly the oxygen tank. As I mentioned, this is really a very difficult thing to make out of carbon fiber, even though carbon fiber has incredible strength to weight. When you then want to put super-cold liquid oxygen and liquid methane, particularly liquid oxygen, in the tank, it is subject to cracking and you have to lay out the carbon fiber in exactly the right way on a huge mold, and you have to cure that mold at temperature. It is just really hard to make large carbon-fiber structures that could do all of those things and carry incredible loads.

In addition to the Raptor engine, the other thing we wanted to focus on was thus the first development tank for the Mars spaceship. This is really the hardest part of the spaceship. The other pieces we have a pretty good handle on, but this was the trickiest one so we wanted to tackle it first. We managed to build the first tank, and the initial test with cryogenic propellant actually looks quite positive. We have not seen any leaks or major issues. This was a massive achievement. Huge congratulations are due to the team that worked on it.

BEYOND MARS

What about going beyond Mars? Generally, I do not like calling things "systems," as everything is a system, including your dog.

However, what we are proposing is actually more than a vehicle; that is the reason we call it a system. There is obviously the rocket booster, the spaceship, the tanker and the propellant plant, and the in situ propellant production.

If you have all four of these system elements, you can go anywhere in the solar system by planet hopping or moon hopping. By establishing a propellant depot in the asteroid belt or on one of the moons of Jupiter, you can make flights from Mars to Jupiter. In fact, even without a propellant depot at Mars, you can do a flyby of Jupiter.

Moreover, by establishing a propellant depot, say on Enceladus or Europa, and then establishing another one on Titan, Saturn's moon, and then perhaps another one further out on Pluto or elsewhere in the solar system, this system really gives you the freedom to go anywhere you want in the greater solar system.

Therefore, you could travel out to the Kuiper Belt, to the Oort cloud. I would not recommend this system for interstellar journeys, but this basic system—provided we have propellant depots along the way—means full access to the entire greater solar system.

Notes

1. This brief introduction is drawn from Roger D. Launius, "Prelude to the Space Age," in John M. Logsdon et al., eds., *Exploring the Unknown: Selected Documents in the History of the U.S. Civil Space Program*, Volume I, Organizing for Exploration, NASA SP-4407 (Washington, DC: Government Printing Office, 1995). Quoted passages are from this essay.

2. For the definitive biography of Wernher von Braun, see Michael Neufeld, *Von Braun: Dreamer of Space, Engineer of War* (New York: Alfred A. Knopf, 2007).

3. This brief introduction is drawn from the author's previous work and from Roger D. Launius, "First Steps into Space: Projects Mercury and Gemini," in John M. Logsdon and Roger Launius, eds., *Exploring the Unknown: Selected Documents in the History of the U.S. Civil Space Program*, Volume VII, Human Spaceflight: Projects Mercury, Gemini, and Apollo, NASA SP-4407 (Washington, DC: Government Printing Office, 2008).

4. Margot Lee Shetterly, *Hidden Figures: The American Dream and the Untold Story of the Black Women Mathematicians Who Helped Win the Space Race* (New York: William Morrow and Company, 2016).

5. This introduction is based on John M. Logsdon, "Project Apollo: Americans to the Moon," in John M. Logsdon and Roger Launius, eds., *Exploring the Unknown: Selected Documents in the History of the U.S. Civil Space Program*, Volume VII, Human Spaceflight: Projects Mercury, Gemini, and Apollo, NASA SP-4407 (Washington, DC: Government Printing Office, 2008).

Ready to find
your next great classic?

Let us help.

Visit prh.com/penguinclassics

PENGUIN
CLASSICS